Manpower and Personnel Needs for a Transformed Naval Force

Committee on Manpower and Personnel Needs for a Transformed Naval Force

Naval Studies Board

Division on Engineering and Physical Sciences

NATIONAL RESEARCH COUNCIL
OF THE NATIONAL ACADEMIES

THE NATIONAL ACADEMIES PRESS
Washington, D.C.
www.nap.edu

THE NATIONAL ACADEMIES PRESS 500 Fifth Street, N.W. Washington, DC 20001

NOTICE: The project that is the subject of this report was approved by the Governing Board of the National Research Council, whose members are drawn from the councils of the National Academy of Sciences, the National Academy of Engineering, and the Institute of Medicine. The members of the committee responsible for the report were chosen for their special competences and with regard for appropriate balance.

This study was supported by Contract No. N00014-05-G-0288, DO #10 between the National Academy of Sciences and the Department of the Navy. Any opinions, findings, conclusions, or recommendations expressed in this publication are those of the author(s) and do not necessarily reflect the views of the organizations or agencies that provided support for the project.

International Standard Book Number-13: 978-0-309-11265-9
International Standard Book Number-10: 0-309-11265-6

Copies of this report are available from:

Naval Studies Board, National Research Council, The Keck Center of the National Academies, 500 Fifth Street, N.W., Room WS904, Washington, DC 20001; and The National Academies Press, 500 Fifth Street, N.W., Lockbox 285, Washington, DC 20055; (800) 624-6242 or (202) 334-3313 (in the Washington metropolitan area); Internet, http://www.nap.edu.

Printed in the United States of America

THE NATIONAL ACADEMIES

Advisers to the Nation on Science, Engineering, and Medicine

The **National Academy of Sciences** is a private, nonprofit, self-perpetuating society of distinguished scholars engaged in scientific and engineering research, dedicated to the furtherance of science and technology and to their use for the general welfare. Upon the authority of the charter granted to it by the Congress in 1863, the Academy has a mandate that requires it to advise the federal government on scientific and technical matters. Dr. Ralph J. Cicerone is president of the National Academy of Sciences.

The **National Academy of Engineering** was established in 1964, under the charter of the National Academy of Sciences, as a parallel organization of outstanding engineers. It is autonomous in its administration and in the selection of its members, sharing with the National Academy of Sciences the responsibility for advising the federal government. The National Academy of Engineering also sponsors engineering programs aimed at meeting national needs, encourages education and research, and recognizes the superior achievements of engineers. Dr. Charles M. Vest is president of the National Academy of Engineering.

The **Institute of Medicine** was established in 1970 by the National Academy of Sciences to secure the services of eminent members of appropriate professions in the examination of policy matters pertaining to the health of the public. The Institute acts under the responsibility given to the National Academy of Sciences by its congressional charter to be an adviser to the federal government and, upon its own initiative, to identify issues of medical care, research, and education. Dr. Harvey V. Fineberg is president of the Institute of Medicine.

The **National Research Council** was organized by the National Academy of Sciences in 1916 to associate the broad community of science and technology with the Academy's purposes of furthering knowledge and advising the federal government. Functioning in accordance with general policies determined by the Academy, the Council has become the principal operating agency of both the National Academy of Sciences and the National Academy of Engineering in providing services to the government, the public, and the scientific and engineering communities. The Council is administered jointly by both Academies and the Institute of Medicine. Dr. Ralph J. Cicerone and Dr. Charles M. Vest are chair and vice chair, respectively, of the National Research Council.

www.national-academies.org

COMMITTEE ON MANPOWER AND PERSONNEL NEEDS FOR A TRANSFORMED NAVAL FORCE

JOHN H. MOXLEY III, Solvang, California, *Co-chair*
CINDY WILLIAMS, Massachusetts Institute of Technology, *Co-chair*
BETH J. ASCH, RAND Corporation
JAMES N. BARON, Yale University
OWEN R. COTE JR., Massachusetts Institute of Technology
LEE F. GUNN, The CNA Corporation
JAMES L. HERDT, Chelsea, Alabama
BARRY M. HOROWITZ, University of Virginia
LEON A. JOHNSON, United Parcel Service
JOHN B. MOONEY JR., Austin, Texas
JUDITH H. MOPSIK, Abt Associates, Incorporated
JOHN E. RHODES, Balboa, California
NANCY T. TIPPINS, Valtera Corporation
JAMES L. WOLBARSHT, DEFCON, Incorporated

Staff

CHARLES F. DRAPER, Director
ARUL MOZHI, Study Director
RAYMOND S. WIDMAYER, Senior Program Officer (as of July 10, 2007)
SUSAN G. CAMPBELL, Administrative Coordinator
MARY G. GORDON, Information Officer
SEKOU O. JACKSON, Senior Project Assistant (as of December 29, 2006)
SIDNEY G. REED JR., Consultant

NAVAL STUDIES BOARD

Staff

CHARLES F. DRAPER, Director
ARUL MOZHI, Senior Program Officer
RAYMOND S. WIDMAYER, Senior Program Officer (as of July 10, 2007)
BILLY M. WILLIAMS, Senior Program Officer (as of August 18, 2007)
EUGENE J. CHOI, Program Officer (through May 18, 2007)
IAN M. CAMERON, Associate Program Officer (through May 21, 2007)
SUSAN G. CAMPBELL, Administrative Coordinator
MARY G. GORDON, Information Officer
SEKOU O. JACKSON, Senior Program Assistant (as of December 29, 2006)

Preface

In December 2005 the Naval Studies Board was briefed by several representatives from the Navy and the Marine Corps, including the Chief of Naval Operations. The discussions during these briefings, which highlighted the pressing manpower and personnel needs of the naval forces in general, were the genesis of the present study on manpower and personnel needs for a transformed naval force.

At the beginning of the 21st century, the Department of Defense (DOD) committed itself to actively transforming the nation's armed forces to meet the military challenges of the future, emphasizing speed and flexibility as well as interoperability among service, joint, and coalition forces. One approach to achieving these goals was by leveraging advances in science and technology. With the onset of the global war on terror, the urgency of that commitment has been accentuated.[1] New technologies and innovations are integral to today's military actions, and associated changes have rippled through all aspects of operations, highlighting the need for changes in policies related to military personnel. The

[1]The National Research Council's (NRC's) Naval Studies Board recently conducted a study on the role of naval forces in the global war on terror (see NRC, 2007, *The Role of Naval Forces in the Global War on Terror: Abbreviated Version*, The National Academies Press, Washington, D.C.). Background information pertaining to the origins of the term "GWOT" can be found in national leadership documents, including (i) The White House, 2006, *The National Security Strategy of the United States of America,* Washington, D.C., March, p. 12; (ii) Office of the Chairman, Joint Chiefs of Staff, 2006, *National Military Strategic Plan for the War on Terrorism*, Joint Chiefs of Staff, Washington, D.C., February 1, p. 3; and (iii) Secretary of Defense, 2006, *Quadrennial Defense Review Report*, Department of Defense, Washington, D.C., February 6. The NRC Committee on Manpower and Personnel Needs for a Transformed Naval Force saw its charter as being neither to endorse nor to replace the term "GWOT."

2006 Quadrennial Defense Review notes that the U.S. military "must continue to adapt to different operating environments, develop new skills and rebalance its capabilities and people if it is to remain prepared for the new challenges of an uncertain future."[2]

The former Chief of Naval Operations (CNO) emphasized the importance of transforming the naval forces to keep pace with progress, pointing out the need to "develop 21st Century leaders . . . through a transformed manpower, personnel, training and education organization that better competes for the talent our country produces and creates the conditions in which the full potential of every man and woman serving our Navy can be achieved."[3] The former Commandant of the Marine Corps (CMC) echoed these sentiments by stating that the Marine Corps "will place renewed emphasis on [its] greatest asset—the individual marine—through improved training and education . . . [and] will continue to attract, recruit and retain the best of America's youth."[4]

As evidenced by the statements of the former CNO and the former CMC, changes in naval personnel practices are necessary in order for the men and women of the Navy and the Marine Corps to continue the process of transformation. Attractive manpower compensation strategies are an essential element of the nation's commitment to the men and women who serve in uniform and thus are crucial to Navy and Marine Corps success in recruiting, retaining, and motivating sailors and marines. Yet the incentive structures created by today's military compensation policies may hamper Navy and Marine Corps efforts to shape their forces to meet the demands of transformation, including network-centric operations.

Military compensation policies also affect transformation in another way. In recent years, the costs of pay and benefits for uniformed personnel have risen dramatically. In the constrained fiscal environment of the future, continued rising personnel costs threaten to drain the Navy and Marine Corps of funding that is otherwise needed to modernize aging equipment, capitalize on emerging technologies such as autonomous systems, and realize the potential of network-centric operations—all of which are essential elements of a transformed naval force.

Recently, there have been several DOD efforts to examine personnel issues in the context of continuing transformation of U.S. military forces. For example,

---

[2]Secretary of Defense (Hon. Donald H. Rumsfeld). 2006. *Quadrennial Defense Review Report*, Department of Defense, Washington, D.C.

[3]Chief of Naval Operations (ADM Michael G. Mullen, USN). 2005. "CNO Guidance for 2006: Meeting the Challenge of a New Era," Department of the Navy, Washington, D.C., October. "Manpower" refers to workforce requirements as defined by the knowledge, skills, and abilities that enable mission accomplishment. "Personnel" refers to the physical asset or resource that the naval services recruit, develop, and manage (shape) to support its identified workforce requirements.

[4]Former Commandant of the Marine Corps (Gen Michael W. Hagee, USMC). 2006. 33rd CMC Updated Guidance: "The 21st Century Marine Corps: Creating Stability in an Unstable World," Headquarters, U.S. Marine Corps, Washington, D.C.

in 2004 the Assistant Secretary of the Navy for Manpower and Reserve Affairs (ASN[M&RA]) issued the *Department of the Navy Human Capital Strategy*[5] aimed at developing a Department of the Navy human resources system that is high-performing, efficient, balanced, aligned, and effective, to achieve the goals for transformation articulated in the Naval Power 21 vision.[6] In 2005, the ASN(M&RA) and the Deputy Assistant Secretary of the Navy, Total Force Transformation (DASN[TFT]) formed a compensation team that drafted strategic goals and guiding principles for future Department of the Navy compensation policy.[7] And recently the Secretary of Defense tasked the Defense Advisory Committee on Military Compensation (DACMC) with identifying cost-effective approaches to military pay and benefits.[8] The DACMC report developed a set of principles for the evaluation of proposed changes to the compensation structure and provided recommendations based on those principles.

TERMS OF REFERENCE

At the request of the former Chief of Naval Operations,[9] a committee under the auspices of the Naval Studies Board of the National Research Council reviewed the military manpower and personnel policies and studies currently underway in the Department of Defense and developed an implementation strategy for the Department of the Navy's future military manpower and personnel needs. The specific tasks were as follows:

- Review the high-priority future staffing needs of the Department of the Navy, taking into account future threats, emerging technologies, and transformational elements such as network-centric operations;
- Review the Department of Defense and the Department of the Navy military manpower and personnel policies and related studies, and summarize

---

[5]William A. Navas Jr., Assistant Secretary of the Navy, Manpower and Reserve Affairs; LtGen Garry L. Parks, USMC, Deputy Commandant, Manpower and Reserve Affairs; and VADM Gerald L. Hoewing, USN, Chief of Naval Personnel. 2004. *Department of the Navy Human Capital Strategy,* Department of the Navy, Washington, D.C., June 21.

[6]Secretary of the Navy (Hon. Gordon England), Chief of Naval Operations (ADM Vern Clark, USN), and Commandant of the Marine Corps (Gen James L. Jones, USMC). 2002. *Naval Power 21 . . . A Naval Vision,* Department of the Navy, Washington, D.C., October.

[7]In addition, the Center for Naval Analyses has provided the Navy with short-term and long-term recommendations for reforming the compensation of naval personnel. See Michael L. Hansen and Martha E. Koopman, 2005, *Military Compensation Reform in the Department of the Navy*, Center for Naval Analyses, Alexandria, Va., December.

[8]Defense Advisory Committee on Military Compensation (ADM Donald L. Pilling, USN [retired], chair). 2006. *The Military Compensation System: Completing the Transition to an All-Volunteer Force,* Arlington, Va., April 28.

[9]ADM M.G. Mullen, USN, CNO, letter dated June 29, 2006, to Dr. Ralph J. Cicerone, President, National Academy of Sciences.

their key recommendations considering the potential benefits and risks of implementation;

- Recommend to the Department of the Navy pilot studies, surveys, policy simulations, experiments, and other mechanisms that can suggest the likely cost, effectiveness, and unanticipated effects of implementing the aforementioned recommendations, and supplement the recommendations where needed (e.g., by exploring career paths, compensation, and so forth);
- Provide strategies to the Department of the Navy for implementation of these recommendations amongst stakeholders, including any organizational considerations for eliminating barriers in the way of the implementation; and
- Determine any new authorities required to implement recommendations as well as potential alternatives to the recommendations that might be made necessary due to impending changes outside the Department of the Navy (e.g., Congress's action to relieve Army personnel stress by raising basic pay).

THE COMMITTEE'S APPROACH

The committee[10] first convened in October 2006 and held additional meetings and site visits over a period of 6 months, both to gather input from the relevant communities and then to discuss the committee's findings and recommendations. Summarized agendas of the meetings are provided below.[11]

- *October 3-4, 2006, in Washington, D.C.* Office of the Chief of Naval Operations, Office of the Assistant Secretary of the Navy for Total Force Transformation, Headquarters U.S. Marine Corps, Office of the Under Secretary of Defense (Personnel and Readiness), and CNA Corporation briefings on military manpower and personnel needs.
- *November 14-15, 2006, in Washington, D.C.* Defense Advisory Committee on Military Compensation, Tenth Quadrennial Review of Military Compensation Committee, Congressional Budget Office, Navy Installations Command, and Office of the Chief of Naval Operations briefings on military compensation and Navy manpower and personnel needs.
- *December 12-13, 2006, in Norfolk, Virginia.* U.S. Joint Forces Command, Fleet Forces Command, U.S. Atlantic Fleet, U.S. Pacific Fleet, Second Fleet, Navy Expeditionary Combat Command, and Naval Network Warfare Command briefings on manpower and personnel needs for a transformed naval force.
- *January 23-24, 2007, in Washington, D.C.* Defense Human Capital Strategy Program Executive Officer, Naval Sea Systems Command, Defense Manpower Data Center, Office of the Chief of Naval Operations, Clemson University, RAND Corporation, Military Officers Association of America, National Military

[10]Biographies of its members are provided in Appendix A.

[11]During the course of its study, the committee held some meetings in which it received (and discussed) materials that are exempt from release under 5 U.S.C. 552 (b).

Family Association, and Fleet Reserve Association briefings on manpower and personnel needs for a transformed naval force, and on military compensation.

- *February 13-14, 2007, in Millington, Tennessee.* Navy Personnel Command, Navy Recruiting Command, Naval Education and Training Command, and Navy Manpower Analysis Center briefings on manpower and personnel needs for a transformed naval force.
- *March 19-23, 2007, in Irvine, California.* Committee deliberations and report drafting.

The months between the committee's last meeting and the publication of the report were spent preparing the draft manuscript, gathering additional information, reviewing and responding to the external review comments, editing the report, and conducting the required security review necessary to produce an unclassified and unrestricted report.

Acknowledgment of Reviewers

This report has been reviewed in draft form by individuals chosen for their diverse perspectives and technical expertise, in accordance with procedures approved by the National Research Council's Report Review Committee. The purpose of this independent review is to provide candid and critical comments that will assist the institution in making its published report as sound as possible and to ensure that the report meets institutional standards for objectivity, evidence, and responsiveness to the study charge. The review comments and draft manuscript remain confidential to protect the integrity of the deliberative process. We wish to thank the following individuals for their review of this report:

Katharine B. Gebbie, National Institute of Standards and Technology,
Stanley A. Horowitz, Institute for Defense Analyses,
Stephen J. Lukasik, Falls Church, Virginia,
Daniel T. Oliver, VADM, USN (retired), Naval Postgraduate School,
Henry P. Osman, LtGen, USMC (retired), Marine Toys for Tots Foundation,
Irene C. Peden, University of Washington,
Jeffrey F. Scott, Philadelphia, Pennsylvania, and
John T. Warner, Clemson University.

Although the reviewers listed above provided many constructive comments and suggestions, they were not asked to endorse the conclusions or recommendations, nor did they see the final draft of the report before its release. The review of this report was overseen by Stephen Berry of the University of Chicago.

Appointed by the National Research Council, he was responsible for making certain that an independent examination of this report was carried out in accordance with institutional procedures and that all review comments were carefully considered. Responsibility for the final content of this report rests entirely with the authoring committee and the institution.

Contents

APPENDIXES

Summary

The men and women who serve in uniform for the naval services are at the heart of U.S. maritime superiority. Sound Navy and Marine Corps manpower and personnel policies are crucial to attracting enough of the right people to join and remain in service for as long as needed. Force management policies must also get people into the right jobs and motivate good performance. Since the inception of the all-volunteer force in 1973, the naval services have had remarkable success in working within the framework of existing policies to attract and retain high-quality people, develop their skills, and manage the force.

The Department of the Navy, however, must not take future success for granted. New threats, emerging technologies, and changing concepts of operations are all combining to place new demands on sailors and marines. Unfortunately, today's manpower and personnel policies will not meet the needs of the Navy and Marine Corps efforts to shape their forces to meet those new demands. At the same time, demographics and other societal forces are acting to reduce the number of military-qualified youths. Moreover, in a constrained fiscal environment, the rising costs of uniformed personnel threaten to drain the Navy and Marine Corps of funding that is needed to modernize aging equipment, capitalize on emerging technologies, and realize the potential for transformation.

As discussed in more detail in Chapter 1, the committee recognizes that the needs of the Navy and the Marine Corps to enable transformation are not congruent. However, the committee also concluded that the best way to address its terms of reference dealing with the Department of the Navy in a single report was to review the challenges to transformation faced by the Department of the Navy and depend on the individual services to move to meet their individual needs. The committee presents analysis and options to meet several potential problems, some facing both services and some only one. The committee's guiding principle was that manpower and personnel changes will inevitably impact both services and thus demand the attention of both.

In 2004 the Department of the Navy published its human capital strategy (HCS)[1] outlining the department's broad goals for human resources. Two years later, in 2006, the Navy's *Strategic Vision: MPT&E [Manpower, Personnel, Training, and Education] Strategic Plan*[2] was published as the first of three milestone documents leading to implementation of the HCS. Later that year the second milestone document, the Navy MPT&E *One Voice Reference Book*,[3] was issued. As of this writing the third milestone contribution—the Navy MPT&E roadmaps containing implementation action plans, discrete tasks required to achieve the HCS, and metrics, accountability methods, and timelines for completion—has yet to be finished. Thus the guidance for implementation of the HCS effort remains unfinished.

Over the course of the past two decades, several studies have recommended fundamental reforms of military manpower and personnel policies across the Department of Defense (DOD). In 2000, the Defense Science Board flatly stated that "the current set of human resource policies and practices will not meet the needs of the 21st century force if left unchanged."[4] In addition, numerous prior studies recommended reforms in the areas of compensation and retirement. In 2005 the Center for Naval Analyses reviewed several past studies to assess the alignment of the Navy's compensation tools with those broad goals.[5] In 2006 the Defense Advisory Committee on Military Compensation (DACMC) called for changes in the basic pay table, consolidation of today's plethora of bonuses and special pays into a few flexible incentives, and an overhaul of the retirement system.[6]

---

[1]William A. Navas Jr., Assistant Secretary of the Navy, Manpower and Reserve Affairs; LtGen Garry L. Parks, USMC, Deputy Commandant, Manpower and Reserve Affairs; and VADM Gerald L. Hoewing, USN, Chief of Naval Personnel. 2004. *Department of the Navy Human Capital Strategy*, Department of the Navy, Washington, D.C., June 21.

[2]Navy Manpower, Personnel, Training, and Education (VADM J.C. Harvey Jr., USN, Deputy Chief of Naval Operations for Manpower, Personnel, Training, and Education, N1). 2006. *Strategic Vision: MPT&E Strategic Plan*, Department of the Navy, Washington, D.C., October 31.

[3]Navy Manpower, Personnel, Training, and Education (VADM J.C. Harvey Jr., USN, Deputy Chief of Naval Operations for Manpower, Personnel, Training, and Education, N1). 2006. *One Voice Reference Book*, Department of the Navy, Washington, D.C., November 1.

[4]Defense Science Board. 2000. *The Defense Science Board Task Force on Human Resources Strategy*, Office of the Under Secretary of Defense for Acquisition, Technology, and Logistics, Washington, D.C., February, p. 14.

[5]Michael L. Hansen and Martha E. Koopman. 2005. *Military Compensation Reform in the Department of the Navy*, Center for Naval Analyses, Alexandria, Va., December.

[6]Defense Advisory Committee on Military Compensation (ADM Donald L. Pilling, USN [retired], chair). 2006. *The Military Compensation System: Completing the Transition to an All-Volunteer Force*, Arlington, Va., April 28.

The committee reviewed the 2005 Center for Naval Analyses report and the 2006 DACMC report. Using an explicit set of criteria based on the needs of a transformed force, the committee critiqued the recommendations of the previous studies. The following findings and recommendations stem from the committee's review of previous studies as well as its understanding of the demands associated with transformation, possible future challenges to the supply of personnel, and developments in the labor force. (The recommendations in the report are not prioritized.)

REALIZATION OF A FULL HUMAN CAPITAL STRATEGY

Finding 1: Completing, communicating, and implementing a comprehensive human capital strategy will be essential to achieving the Navy and Marine Corps transformation goals.

In 2004 the Department of the Navy published an overview document outlining an HCS. The committee is concerned that the strategy remains incomplete, in that the declared third step, the implementing roadmaps, have yet to be issued by the Navy. Nor is the Department of the Navy HCS broadly understood or routinely employed as a guide to decision making outside a relatively narrow circle of manpower and personnel specialists. Moreover, since the signing of the Navy's HCS in June 2004, the services have undertaken important operational and organizational changes. They have also gained experience related to the Navy's downsizing, lengthy commitments in Iraq and Afghanistan, and increased reliance on the private sector—all without the coherence and logic that a completed HCS would bring to the decision-making process.

Recommendation 1: The Chief of Naval Operations (CNO) and the Commandant of the Marine Corps (CMC) should take ownership of their services' human capital strategy (HCS) and direct its prompt completion. Beyond that, the CNO and the CMC should institute a process to review and update their HCS in light of changes in the strategic environment, future plans, and evolving experience with existing human resource policies. The completion of the services' HCS should be done with the following criteria in mind:

1. *Aligned.* The HCS should be linked clearly to the services' goals and missions, identifying the highest-priority "key success factors" required of personnel for organizational success.
2. *Internalized.* The HCS must be communicated to and broadly understood at all levels, in ways that clarify to individuals in each subunit how their efforts affect overall success.

3. *Routinized.* The HCS should routinely inform decisions, trade-offs, and resource allocations and should be embedded in everyday operating procedures (e.g., planning and budgeting, personnel reviews, external reporting).

4. *Coherent.* The HCS should promote coherence and synergies in human resource administration across specific domains (e.g., recruitment, compensation, training, and development). It should sustain a human resources "brand" that makes clear to current and prospective sailors and marines what is expected of them and what they in turn can expect of the organization.

5. *Measurable.* The HCS should describe desired outcomes that can be and are assessed with metrics.

6. *Adaptable.* The HCS should be dynamic, undergoing routine reassessment and adjustment in light of learning and of changing organizational and environmental contingencies.

7. *Consequential.* Supporting the HCS should represent (and must be *perceived* to represent) a significant element in the formal assessment and evaluation of leaders.

The committee recommends an assessment of current proposed human resource strategies against this list of criteria and creation of a template simple enough in form and content that it can be used to articulate the key success factors and human resource strategy to diverse audiences at all levels of the naval services.

RECRUITMENT OF SAILORS AND MARINES

Finding 2: Changes in demographics, fitness, and attitudes toward military service call for creative approaches to Navy and Marine Corps recruiting.

Since the inception of the all-volunteer force, the Navy and Marine Corps have devoted substantial resources to recruiting. Those resources have been instrumental in bringing sufficient numbers of high-quality youth into the services over the years and making the all-volunteer force a success.

In the future, military-qualified youth will make up a shrinking share of the U.S. population. Attitudes toward serving in the military may also reduce the pool of potential recruits. At the same time, the nation's racial and ethnic mix is shifting in ways that will advantage employers that emphasize and value diversity and make extra efforts to attract minorities.

Reserve Officer Training Corps (ROTC) scholarships are a crucial tool in attracting qualified students from diverse backgrounds into the officer corps. Complementing that effort, the Junior ROTC offers opportunities for high school students, including many who are otherwise disadvantaged, to improve their physical fitness, build values, and get a taste of military life.

Recommendation 2: The Assistant Secretary of the Navy for Manpower and Reserve Affairs, the Chief of Naval Personnel, and the Deputy Commandant of the Marine Corps for Manpower and Reserve Affairs should examine options to expand Junior ROTC programs to attract qualified students from diverse backgrounds to naval service.

Congress would have to be persuaded to provide additional funds for Junior ROTC programs. In some cases funds have been appropriated but have not been spent by the services. Location of Junior ROTC units can be critical to the success of the program. School districts with a large minority population need to be emphasized for the new units. Furthermore, there are aviation magnet high schools in the United States that should be made prime targets for units. Under current law only retired active duty members who are drawing retirement can be instructors in Junior ROTC. The way pay is structured, the retired reservist who is not drawing retirement cannot be an instructor. The Secretary of the Navy needs to determine if this requirement can be changed to increase the pool to draw on for instructors.

DEPLOYMENT AND PACE OF WORK

Finding 3: Recent and planned changes in Navy and Marine Corps missions, unit structure, and materiel will likely have adverse effects on people and readiness.

Studies based on surveys and focus groups indicate that whether service members are deployed or at home, their morale and intention to continue in service appear to suffer when work hours are unexpectedly onerous, and when the realities of military life differ substantially from the individuals' expectations. The wars in Iraq and Afghanistan, other worldwide deployments and commitments including the global war on terror, and the continual downsizing of the Navy pose the potential for adverse effects on people and readiness. While such effects may ultimately be ameliorated by the introduction of labor-saving technologies, they are exacerbated when workloads increase because the fielding of or training on new equipment lags behind the personnel reductions. Moreover, the challenges of irregular warfare, the continued shifting of shore work to contractors, and the Navy's plans to move from a pyramidal to an oval workforce structure (discussed in Chapter 2) are likely to lengthen work days for many sailors and marines. If individuals are overloaded beyond their ability to support the mission, both mission readiness and future retention are likely to suffer.

Recommendation 3: The Navy and Marine Corps need better evaluation programs to measure and interpret the effects of changes in workload, separation from home, length and repetition of deployments, and similar factors on readiness, morale, and intention to serve. The Assistant Secretary of the Navy for Manpower and Reserve Affairs, the Chief of Naval Personnel, and the Deputy Commandant of the Marine Corps for Manpower and Reserve Affairs should develop new metrics for the early identification of morale and retention signals related to deployment cycles, workload, and manning levels. Additional studies should be undertaken to understand better how deployment cycles, workload, morale, intention to continue, and actual continuation are related.

CAREER PATHS

Finding 4: Officer career paths in the Navy and Marine Corps are designed to produce well-rounded officers suited for major command and flag rank. However, relatively few officers achieve these goals; most of the career force leave at (or shortly after) 20 years—many at the peak of their operational experience and expertise.

Recommendation 4: The Secretary of the Navy (SECNAV), the Chief of Naval Operations (CNO), and the Commandant of the Marine Corps (CMC) should take steps to allow those officers who prefer an operational or specialist career to remain in operational or specialist billets, with limitations on rank, for their entire careers. Toward that end the Navy Department leaders should investigate how to embed performance incentives to accommodate career paths that would not involve moving up the chain of command. In addition, the SECNAV, CNO, and CMC should sponsor and support changes to the basic pay table that would allow individuals to stay in grade longer without financial penalty.

Finding 5: Current promotion patterns from enlisted to commissioned officer are wasteful of talent and inconsistent with the Navy's desire to move toward an "oval workforce." The Navy recruits more ensigns than it needs in order to ensure that it will have enough midlevel officers. The surplus of ensigns is inconsistent with the Navy's desire to develop for its future ships a workforce that looks less like a pyramid and more like an oval. At the same time, the Navy provides limited advancement opportunities for E-8s and E-9s, who therefore leave the service in large numbers as soon as or soon after they reach retirement age.

Recommendation 5: The Chief of Naval Operations should work with the Secretary of the Navy to institute an enlisted-to-commissioned promotion path that pulls senior enlisted people into the midlevel officer ranks, and should use this program to avoid recruiting more ensigns than needed.

Finding 6: Today's restrictions on entry levels for people will hamper Navy and Marine Corps efforts to exploit emerging technologies and address unexpected threats.

To exploit fully the trend from hardware- to software-driven systems, the Navy and Marine Corps will need officers and enlisted personnel who possess information technology skills at a very high level. Such skilled individuals will have to provide the Department of the Navy with educated buyers and allow units to modify their systems at the pace the threat requires, rather than the pace the defense industry normally provides. These skills are increasingly in demand in the private sector as well.

When a new requirement suddenly emerges, such as the cultural and linguistic knowledge demanded by the global war on terror, there is no time to develop the expertise from within. Whether or not such expertise will be needed over the long term, the short-term need is often urgent. Individuals with such skills are available in the civilian world. If provided the right incentives, such individuals might volunteer to serve on active duty for a specified period. Such individuals would be more valuable to the Department of the Navy than civilian contractors, and such service might be more attractive for the individuals as well.

Recommendation 6: The Chief of Naval Operations and the Commandant of the Marine Corps should direct the development of a plan for lateral-entry programs to permit and encourage routine entry from the civilian world into active duty at all ranks for individuals with needed skills.

Finding 7: Today's policies can make a return to active-duty service challenging or unattractive for sailors and marines who wish to take time out to obtain training and education not offered through the services, gain experience in the civilian world, or start a family.

Today some sailors and marines of both genders leave for educational or workplace opportunities that are not available to them on active duty. Women early in their careers exit the services in disproportionate numbers. While they may need only a few years off, those who leave generally do not return to service. The early exit of experienced sailors and marines puts an extra burden on the remainder of the force and on resources for recruiting and training.

Recommendation 7: The Chief of Naval Operations and the Commandant of the Marine Corps should direct the development of programs for on/off ramps whereby sailors and marines could without undue prejudice to their careers take time off from their active-duty careers in order to obtain education, take advantage of training opportunities beyond those provided by the Navy and Marine Corps, or start a family.

COMPENSATION AND RETIREMENT REFORM

Finding 8: The current package of immediate compensation is overly complex, lacks flexibility, is not conducive to Navy and Marine Corps force management, and generally costs more than its value to service members.

Recommendation 8: The Secretary of the Navy, the Chief of Naval Operations (CNO), and the Commandant of the Marine Corps (CMC) should support the following DOD-wide policy changes:

- Improve incentives for performance by reconfiguring the basic pay table to set pay based on grade and time in grade rather than time in service.
- Equalize the basic allowance for housing (BAH) rate for all service members, regardless of family status.
- Pay BAH to all service members and charge rent to those in government housing, with the rent equal to the fair market value of their housing.
- Consolidate all deployment-related pays, including the family separation allowance. Deployment-related pay should be set to reflect the nature of deployment.
- Consolidate the special and incentive pays into a much smaller number of categories, and work actively to help shape the categories to make these pays as flexible and useful as possible. In doing this, give particular attention to assignment and career incentive pay and selective reenlistment bonuses to best match the interests of sailors and marines with those of the services. Assuming that the broader pays are instituted, naval leaders, including personnel and budget offices, should make every effort to avoid "cost creep" as individual constituencies lobby for pay increases.

The CNO and CMC should direct that rigorous evaluations of the services' quality-of-life programs be carried out to understand their effects on readiness and ensure their cost-effectiveness. The leaders should also conduct research to assess the costs and benefits of flexible spending accounts for these programs, including health care.

Finding 9: The current military retirement system impedes flexible force management and is inequitable and inefficient. In addition, health care benefits for military retirees are not cost-effective.

Recommendation 9: The Secretary of the Navy (SECNAV), the Chief of Naval Operations (CNO), and the Commandant of the Marine Corps (CMC) should advocate and support changes to the compensation system that would accomplish the following:

- Encourage longer careers for some personnel and shorter careers for others.
- Offer greater flexibility in management of active and reserve forces.
- Result in more personnel who are vested.
- Generally provide more compensation up-front and less deferred to retirement.
- Improve integration of active and reserve forces.

A three-part system seems the most promising in terms of effectiveness and political feasibility. Such a system would consist of the following:

- An entitlement that allows members to accumulate benefits for old age,
- A career management tool for the services, and
- Additional front-loaded compensation that would also serve as a management tool.

The SECNAV, CNO, and CMC should also support a proposal that would require retirees under age 65 to stipulate a primary and secondary insurer, and that would give these retirees an incentive to stipulate Tricare as the secondary insurer. To induce more cost-effective use of health benefits, Tricare fees should be indexed to the annual cost-of-living adjustment to the military retirement annuity. Finally, the financing of the under-65 health benefit should be through an accrual fund, as it is for retirees over 65.

RESEARCH TOOLS AND EXPERIMENTATION

Finding 10: Establishing and sustaining transformation in Navy and Marine Corps manpower and personnel policies will require the use of a variety of research tools. Research tools ranging from surveys and analyses of administrative data to pilot demonstrations and experimentation have always been key tools in formulating policies that support military manpower transformation. Experimentation has been one of the research tools that the Department of Defense and the Navy in particular have used in achieving changes to the shape of military forces, and the Navy's recent Sea Swap project illustrates that the Navy understands the value and challenges of experimentation.

Recommendation 10: The Assistant Secretary of the Navy for Manpower and Reserve Affairs, the Chief of Naval Personnel, and the Deputy Commandant of the Marine Corps for Manpower and Reserve Affairs should continue to develop and use research tools and experimentation to address the many facets of the naval services' transformation equation. In particular, the committee recommends the following projects:

- Evaluate Navy plans for dealing with the special requirements imposed by the crewing needs of the littoral combat ship (LCS) and other ship classes by using "embedded ovals," that is, subsets of existing combatant crews, assigned to existing ships, to simulate the crews of LCSs in service.
- Design and conduct a controlled experiment to test the assumptions of the value of various leave options, including the notion of the on/off ramp.
- Undertake a pilot demonstration of a lateral entry and exit program to evaluate the availability of needed talent outside the services, the willingness of people with such talent to serve for a limited duration, and the contribution that such people could bring to real-world naval service situations.
- Use the Navy's Job Advertising and Selection System, Super-JASS, to create an experiment in the allocation of sailors to sea duty by using a variable economic incentive that could replace the blanket amount of sea pay now in use.
- Use administrative data and other research tools to identify warfare behaviors and service appropriate to the requirements of irregular warfare, and encourage such behaviors and service by publicizing and rewarding them.

CHANGE MANAGEMENT

Finding 11: The implementation of changes to manpower and personnel policies is a complex and difficult process.

The SECNAV, CNO, and CMC already have the authorities they need to make some changes. Even in those cases, however, change may be complex or involve several stakeholders, including the individuals who serve, their families, military retirees, the other military services, and Congress. If manpower and personnel policies are going to meet the challenges of the future, the Department of the Navy must prepare for and overcome strong opposition.

Recommendation 11: To bolster the likelihood that changes to manpower and personnel policies will be adopted, the Secretary of the Navy, working with the Chief of Naval Operations and the Commandant of the Marine Corps, should take control of the change management processes. The Department of the Navy-wide processes should include the following:

- Opportunities for early successes,
- A program of continuing evaluation, and
- The institutionalization of a program of comprehensive education of and communication with service members and other stakeholders about the reasons for the changes to manpower and personnel policies and the desired outcomes of change.

1

Introduction

It has long been recognized that the ultimate key to success of U.S. naval forces is the quality of their people. A highly trained, versatile force is critical to realizing the full potential of our naval forces. In his 2006 guidance, ADM Michael G. Mullen, USN (Chief of Naval Operations [CNO]) states, "Our success in defense of this nation depends upon the men and women of the United States Navy—active, reserve, and civilian and their families."[1] His guidance sets as a priority "Develop 21st Century Leaders . . . through a transformed manpower, personnel and training organization that better competes for the talent our country produces and creates the conditions in which the full potential serving our Navy can be achieved."[2]

Gen James T. Conway, USMC, Commandant of the Marine Corps (CMC) pointed out in his recent guidance: "Our marines and sailors in combat are our number one priority in all that we do."[3] The Commandant's guidance also addresses the current force and the global war on terror (GWOT), and he provides direction to determine the structure and manning requirements to meet the Marine Corps' goal of ensuring that individuals and units have 12 months at home for every 6 months they are deployed: "Deputy Commandants and Headquarters Marine Corps Directors will . . . examine our requirements for recruiters, infra-

[1]ADM Michael G. Mullen, USN, Chief of Naval Operations. 2005. "CNO Guidance for 2006: Meeting the Challenge of a New Era," Department of the Navy, Washington, D.C., October, p. 2.

[2]ADM Michael G. Mullen, USN, Chief of Naval Operations. 2005. "CNO Guidance for 2006: Meeting the Challenge of a New Era," Department of the Navy, Washington, D.C., October, p. 3.

[3]Gen James T. Conway, USMC, 34th Commandant of the Marine Corps. 2006. "Commandant's Planning Guide," Headquarters, U.S. Marine Corps, Washington, D.C., p. 1.

structure, materiel, and equipment to support both manning at a 1:2 deployment to dwell ratio and training across the spectrum of warfare."[4]

The committee recognizes that overall naval manpower includes civilians who work for the Department of Defense (DOD) and defense contractors as well as uniformed men and women. However, even though civilians and contractors perform very important tasks, this report focuses on men and women in uniform.

The committee also recognizes that the transformational needs of the Navy and the Marine Corps are not identical. The Navy is faced with integrating a new generation of ships demanding smaller and more highly trained and technically proficient crews than the legacy fleet. The Navy must also move quickly to broaden its mission from the traditional major combat operations to incorporate the littorals and major estuaries—green- and brown-water operations. All of this while maintaining the readiness of the legacy ships as they are retired over the next 30 years.

The Marine Corps, on the other hand, faces a less dramatic need for change in the short term. Their primary mission as an initial amphibious strike force will continue. In the longer term the situation is not as clear. In Iraq and Afghanistan the Marine Corps and Army missions are not clearly separated. The role of the Marine Corps vis-à-vis the expanding role of the Special Forces of the Navy and the Army is also not clearly defined. Beyond that the role of the developing Marine Corps Special Operations Forces is in the process of being defined.

Both the Navy and the Marine Corps will face recruitment and training challenges and compensation issues as they increasingly compete with the private sector for talent. Here the Marine Corps mandate to expand will heighten the need for numbers whereas the reduced crew sizes will ease the need of the Navy to meet end-strength targets.

Rather than separate each of the issues into Navy and Marine Corps compartments, the committee chose to lay out the spectrum of change issues and depend upon the leadership of each service to determine how each issue challenges them and how to respond.

The committee heard from key leaders, managers, and stakeholders about dramatic changes in operations, technology, equipment, and force structure that are ongoing or planned in the Navy and Marine Corps and their implications for manpower and personnel. The committee also reviewed several previous studies for their insights into the manpower and personnel challenges that face the Navy and Marine Corps and possible ways to address those challenges.

Based on these multiple sources of information, the committee concluded that despite existing leadership guidance, the Navy and Marine Corps have not yet come to grips with the urgency, importance, and difficulty of building manpower

[4]Gen James T. Conway, USMC, 34th Commandant of the Marine Corps. 2006. "Commandant's Planning Guide," Headquarters, U.S. Marine Corps, Washington, D.C., p. 5.

and personnel policies and systems that will allow the services to get the right people into the right jobs and provide incentives for productivity. Too much of the manpower and personnel structure today still rests on the notion that people are largely interchangeable. When problems surface in immediate crises, temporary solutions are often patched together quickly, leaving behind a complex patchwork of policies without addressing the underlying causes.

The Department of the Navy is at a critical crossroads. It can continue to risk the consequences of the traditional approach, which often assumes that the goal of manpower and personnel policies is simply to meet a given force level. If that is the case, focusing on recruitment and retention will suffice. If, on the other hand, the Department of the Navy embraces transformation and the manpower changes that it demands, it must fully accept and implement the recent guidance of the CNO and the CMC. The naval forces must develop new manpower policy and implementation plans.

Many of the manpower and personnel problems that the naval services face are deep; solving them will require time and the cooperation of key stakeholders inside and outside the department. Problems in this category include those caused by the retirement system and the time-in-service pay table. For such problems there will be no quick fixes. Nevertheless, the committee believes that some changes—for example, the completion of the human capital strategy, the development of better evaluation programs to assess the impact of repeated deployments, and experimentation to assess the implications of future manning of the littoral combat ship (LCS)—can be made relatively quickly, and that the CNO and CMC are in a position to make them. Such changes could result in important early improvements that the Navy and Marine Corps could use as leverage for the more difficult steps.

For the Navy and Marine Corps to bring their ambitious plans for the future to fruition, the CNO and CMC must make manpower transformation a very high priority. They must personally lead and become identified with the manpower transformation process. They must be personally involved at every step along the way. In addition, the Department of the Navy leaders must involve every sailor and marine in the transformation process and ensure that they understand and embrace it.

The road will be difficult and involve integrated changes to virtually every aspect of manpower and personnel policy—including recruitment, career paths, force management, compensation, retirement, education, and training. But, once these changes are in place, the naval forces will be prepared to meet the future. To take an easier path will mean a future of constantly cobbling together temporary solutions to chronic problems, much as is done today.

The next section of this chapter discusses how past policies still shape conditions of the present. The chapter goes on to describe the broad changes envisioned for the naval forces and their implications for men and women in uniform and the policies that surround them. The chapter ends with a brief outline of the report.

TODAY'S POLICIES REFLECT PAST NEEDS

To gain insight into origins of the manpower situation facing today's naval forces, one must go back centuries in time to the beginning of naval missions and manpower policy. For our purposes, however, a return to the World War II (WWII) era will suffice. In WWII the United States successfully undertook a massive mobilization of conscripted forces that led to one of the most significant military victories in history. The enormity of the effort, however, was followed by major manpower and demobilization issues in the postwar period. In an effort to resolve the problems, the Congress passed the National Security Act of 1947. Much of this act is still enshrined in Title 10 of the U.S. Code, which governs many aspects of the armed forces.

One of the most important revisions to Title 10 came with the Defense Officer Personnel Management Act (DOPMA) of 1981 and a revised DOPMA in 1984. The two DOPMAs set limits on the proportion of officers in each service who can hold the rank of Major (Lieutenant Commander for the Navy) or higher at any one time. They also stipulate "up-or-out," or high year-of-tenure rules for officers that are consistent across all four services. During its information gathering sessions, the committee heard repeatedly from Navy and Marine Corps officials that the DOPMAs' constraints on the proportion of senior officers and its up-or-out rules hamper the achievement of future force goals.

Two other laws that constrain the shape of the force and the flexibility of individual career paths are the Warrant Officer Management Act (WOMA), adopted through the Fiscal Year 1992/1993 National Defense Authorization Act, and the Reserve Officer Personnel Management Act (ROPMA), which became effective on October 1, 1996. WOMA established a single promotion structure for warrant officers regardless of service, established the rank of CW-5, stipulated high year-of-tenure rules for the warrant officer ranks, and authorized the service secretaries to implement mandatory retirement for selected warrant officers who are eligible to retire. ROPMA streamlined promotion guidelines and limited overall promotions, thereby establishing rules that more closely resemble the active-duty promotion blueprint. The act also addressed mandatory separation dates and officer retention based on the specific needs of the service.

Despite important events and changes since 1947, the Korean Conflict, the Cold War, the Vietnam War, the introduction of the all-volunteer force, the stunning changes in technology, and now the GWOT, overall manpower policies established in 1947 remain largely intact. The basics of the system of promotion for officers and the principles of military compensation and retirement have changed very little.

FUTURE OPPORTUNITIES AND CHALLENGES

Yet the missions of naval forces, particularly the fighting missions, have changed significantly since WWII and the Cold War. In its fighting missions the

U.S. Navy, particularly the submarine force, underwent a period of intense innovation during the 1950s and early 1960s. Later decades of the Cold War saw little change, however. Had the Cold War erupted into a major battle on the plains of Europe, the Navy's war missions were to destroy the Soviet Navy, particularly its submarines, defend convoys in the Atlantic, and perhaps, in concert with the marines, conduct strike operations on the periphery of the Soviet Union. These large-scale, blue-water operations would not have been greatly dissimilar from those launched during WWII.

The post-Cold War changes were described in some detail in a recent Congressional Budget Office (CBO) report.[5] According to that CBO report, the major peacetime change has been the addition of more operations to provide homeland security in U.S. coastal waters, a mission shared with the U.S. Coast Guard. In addition, naval resources are being called upon for humanitarian missions in the wake of natural disasters, such as tsunamis and major hurricanes. And there is the need to win the support of regional populations in the crucial campaign for hearts and minds. For the Navy that means, in addition to specific humanitarian missions, an increase in port calls, military-to-military operations, and engagement through foreign area officers.

Naval war missions have changed to include large and small strike missions using land-attack missiles and carrier-based aircraft, the sinking of an occasional small enemy ship, and antimine operations. The use of unmanned aerial vehicles as persistent surveillance platforms is growing, as is the demand for special forces. Day-to-day patrol and maritime intercept operations in waters like the Persian Gulf and around the Horn of Africa are in great demand. Visit, boarding, search, and seizure (VBSS) is a mission requirement imposed now on every Navy ship, regardless of the size of its crew.

Now during the GWOT, the Navy faces a new type of adversary who avoids direct confrontation, operates in small groups, and attacks primarily civilian targets. Iraq's use of mines during the Persian Gulf War of 1991 suggests that countries with even moderate military capabilities can pose a challenge when they use their modest sea-denial capabilities in innovative ways.

Furthermore, the pace of technical change continues. Information technology, which plays a growing role in modern military operations, is also a major component of an increasingly globalized commercial economy. If the United States faces a peer competitor in a future war, it will be one that possesses technological capability more comparable to that of our own forces than did the Soviet Union. (Chapter 2 provides a more complete discussion of new threats.)

The central ground fighting role of the Marine Corps in both Operation Enduring Freedom and Operation Iraqi Freedom speak to the new challenges facing that service. Less obvious today, but equally challenging, will be the new

---

[5]Congressional Budget Office. 2006. *Recruiting, Retention, and Future Levels of Military Personnel,* Washington, D.C., October.

concepts of developing the ability to maneuver from the sea by passing over, rather than through, opposing coastal defenses.

These new naval missions do not replace more traditional ones. The Navy will retain requirements for ensuring freedom of the seas against more traditional threats, which have not disappeared.

In summary, although basic manpower policy has changed relatively little, the missions of naval forces have changed significantly. Large blue-water confrontations seem unlikely in the near term. They have been replaced by significant responsibilities in the littorals of the world and in homeland security. These new responsibilities require more widely distributed forces that are capable of independent action and able to carry out diverse functions that depend on the particular scenario.

MANPOWER CONCERNS FOR NAVAL FORCES

Despite guidance, quadrennial reviews of compensation, and numerous study group recommendations, the maritime services still face major manpower issues in preparing for a transformed naval force. Current plans call for the number of sailors to continue to decline, even as the number of ships increases (Figure 1.1). In contrast, the Marine Corps will expand in size during the coming 6 years (Table 1.1). To meet requirements, U.S. naval forces face challenges in achieving manpower goals. The challenges include personnel numbers, both in the aggregate

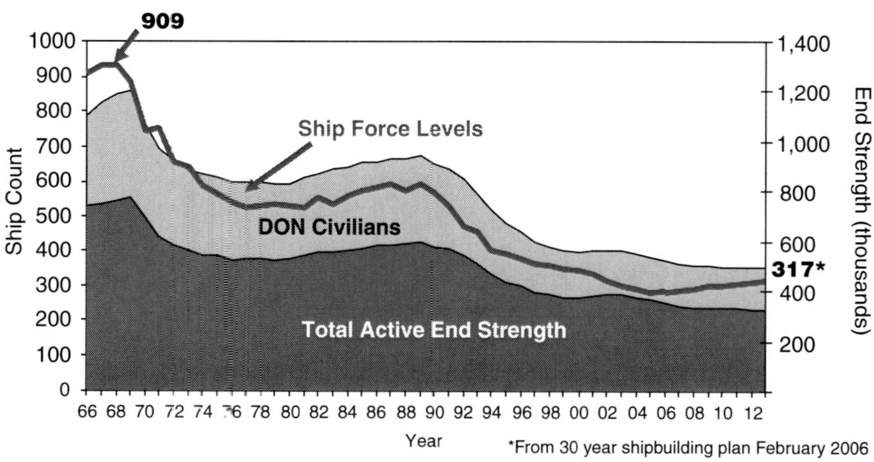

FIGURE 1.1 Number of sailors will decline as number of ships grows. SOURCE: VADM John C. Harvey Jr., USN, Deputy Chief of Naval Operations for Manpower, Personnel, Training, and Education and Chief of Naval Personnel, "Shaping the Force for the Future," presentation to the committee, October 3, 2006, Washington, D.C.

TABLE 1.1 Current and Planned End Strength of the Naval Services

	FY 2006 Authorized	FY 2013
Navy		
Active duty	352,700	322,000
Reserve	73,100	73,100
Marine Corps		
Active duty	179,000	202,000
Reserve	39,600	39,600

SOURCE: Congressional Budget Office, 2006, *Recruiting, Retention, and Future Levels of Military Personnel,* Washington, D.C., October, pp. 42, 54; RADM David A. Gove, USN, Deputy Chief of Naval Personnel, Commander, Navy Personnel Command, "Manpower and Personnel Needs for a Transformed Naval Force," presentation to the committee, February 13, 2007, Millington, Tenn.; DOD press release, "Fiscal 2008 Department of Defense Budget Released," February 5, 2007.

and in specific occupational categories, shifts in the demographics of the U.S. population, and a decline in the propensity of young men and women to serve in the military. Changes in the mix of skills will be required of officers and enlisted personnel in both the active-duty and reserve forces. The services also face dramatic changes in expectations among reservists and increasing competition from the private sector for human capital.

Transformation goals call for developing smaller, more nimble, technologically advanced units of power networked into a mobile lethal force. There is concern that fielding the highly trained, versatile force necessary to realize the full potential of the technological innovation is lagging. Until recently the Navy and Marine Corps generally took for granted that the manpower needed for a transformed force would be readily available and could be trained to meet the new demands. Several factors call this assumption into question. During the committee discussions, Navy leadership has expressed deep concern over the manpower issue and the lack of attention to it.

Challenges for Recruiting

All of the services are challenged to meet their numerical recruiting goals while maintaining a high-quality force. The Navy's recruiting goals for 2007 are 10,600 reserve enlisted, 37,000 active enlisted, and 1,570 active-duty officers. Currently the Navy has been able to meet its numerical goals while adhering to its quality standards, which are tighter than those set by law and regulation for the DOD overall.[6] While these recruiting standards and goals are currently

[6]The Navy expects 95 percent of its enlistees to be high school diploma graduates and at least 70 percent to score above the median on the military's entrance test of cognitive aptitude, the Armed Forces Qualification Test (AFQT). Unlike the Army in recent years, the Navy will not accept any

attainable, sustaining quality and quantity in the future will likely require greater effort and expense.

As is discussed in Chapter 2, the number of qualified prospective male recruits is declining for several reasons. The propensity to enlist is lower now than at any other time in recent history. The continuing war in Iraq is changing young people's perceptions about military service. The share of immigrants in the population places an additional constraint on recruiting, as a large number of immigrants do not meet the services' standards for high school graduation or English language [7]

Key influencers are less likely to advise young people to serve in the military. A declining share of today's parents and grandparents have served in the military, and therefore the role model of military service provided by one's father or grandfather is not as prevalent as it was even a decade ago. Public opinion against the war in Iraq has also taken its toll on support for joining the military among those who influence young people's decisions, including parents, teachers, and coaches.[8]

Perhaps most troubling is the declining share of young people who meet the services' physical, mental, health, and moral standards. In 2005 some 72 percent of young men were unqualified for service for one reason or another, cutting deeply into the pool from which the Navy and Marine Corps can recruit troops.[9] Particularly troubling in recent years is the large fraction of youth who are disqualified because of obesity.

Recruitment challenges will be compounded during the coming years. In the Marine Corps the service's planned expansion will require more recruits each year. For the Navy the challenge will be getting the right quality of the recruits for new categories of billets related to the GWOT and to move from a pyramidal to an oval workforce structure.

To address these challenges, the Navy is increasing its recruiting force and is considering options to relax some of its quality goals, as well as offering several educational incentives for potential recruits. The Marine Corps plans to expand its recruiting force, but hopes to maintain current levels of quality. The Marine Corps is considering increases in recruiter bonuses and advertising budgets.

---

recruits whose score on the AFQT falls at or below the 30th percentile (category IV). See RADM(S) Joseph F. Kilkenny, USN, Commander, Navy Recruiting Command, "Navy Recruiting Command . . . Seeking the 'Best' and the 'Brightest,'" presentation to the committee, February 13, 2007, Millington, Tenn.

[7]See, for example, the Pew Hispanic Center tabulations of the 2005 American Community Survey. Also see the U.S. Census Bureau, U.S. Census of Population, 1960, 1970, and 1990; available at <http://www.census.gov/population/www/socdemo/educ-attn.html>. Accessed on September 17, 2007.

[8]Kristin Roberts. 2007. "Military Sees Parents as Big Recruiting Barrier," Reuters, May 11.

[9]RADM(S) Joseph F. Kilkenny, USN, Commander, Navy Recruiting Command, "Navy Recruiting Command . . . Seeking the 'Best' and the 'Brightest,'" presentation to the committee, February 13, 2007, Millington, Tenn.

These steps will clearly help the services meet their recruiting goals in the short term, but it is not at all clear that they will suffice in the longer term. Additional efforts to expand the pool of qualified and interested applicants may be needed to solve the longer-term problems of meeting challenging recruiting goals.[10]

Transformation Requires Investment in People

The military's weak record of confronting manpower and personnel issues head-on has been reinforced over the years by the conventional wisdom that manpower was relatively inexpensive compared with technology. Future capabilities and technology, however, will require individuals to master multiple skill sets and the naval forces to provide increased and continuing training and education for all hands. Recently Navy Secretary Donald Winter stated that "we are increasingly seeing that tactics, techniques and procedures—and the people who utilize them—are becoming more important than the technology itself." He went on to say, "What will win this war [global war on terror, or GWOT] is the human factor."[11]

To generate the enhanced manpower capability will require an investment in people, including recruitment, training, education, retention, and separation of appropriate individuals. Human-systems integration, the careful design of man-machine (technology) systems, will also be increasingly important for ensuring successful human performance. Transformation is characterized by both technical and procedural innovation. To master the capabilities necessary to effectively implement transformation will almost certainly require most personnel to master several skill sets. Such mastery will require ongoing training and education throughout a service member's entire career.

An example of the need for continuing education is the CMC initiative for the Center for Advanced Operational Cultural Learning (CAOCL) and the requirement for all career marines (enlisted before reaching the grade of E-7 and officers before reaching O-4) to gain operationally relevant regional, cultural, and language knowledge to allow them to plan and operate successfully in the joint and combined expeditionary environment in any region of the world.[12] Such investments in training and education are costly, but they are necessary to ensure a force capable of meeting the challenges of the new missions.

[10]See Section 9528 of the "No Child Left Behind" legislation, 115 Stat.1425 Public Law 107-110, January 8, 2002, 107th Congress, as an example of creating and retaining access for armed forces recruiters to high school juniors and seniors.

[11]Remarks by Donald C. Winter, Secretary of the Navy, Surface Navy Association National Symposium, Crystal City, Va., January 9, 2007.

[12]Col Otto J. Rutt, USMC, Senior Marine Advisor to the Defense Human Capital Strategy Program Executive Officer (DHCS PEO), "21st Century Marine Corps," presentation to the committee, January 23, 2007, Washington, D.C.

Young people with the learning abilities for multiple, complex tasks will be in great demand by all employers, military and civilian alike. Once highly trained, these sailors and marines will possess very marketable skills and be even more attractive to the civilian sector. It will require considerable effort for the naval forces to attract and retain the manpower required to carry out their missions.

The Cost of Pay and Benefits Is Rising

The new manpower needs come at a time when military pay and benefits consume about one-third of the DOD budget and place considerable pressure on accounts for acquisition and for operations and maintenance. The proportion of budgets devoted to military pay and benefits is substantially higher for the Marine Corps, both because that service is more people-intensive and because most of the Marine Corps' aircraft, as well as the ships from which they operate, are purchased by the Navy. The Navy's costs for military pay and benefits too will continue to rise even as the size of the force shrinks (Figure 1.2), fueled significantly by the rising cost of benefits for retirees. It is also apparent that, at least in the near term, naval manpower costs are likely to continue to rise more rapidly than the overall Department of the Navy budget for similar reasons—the rising cost of health care benefits and recent increases in both pay and fringe benefits.

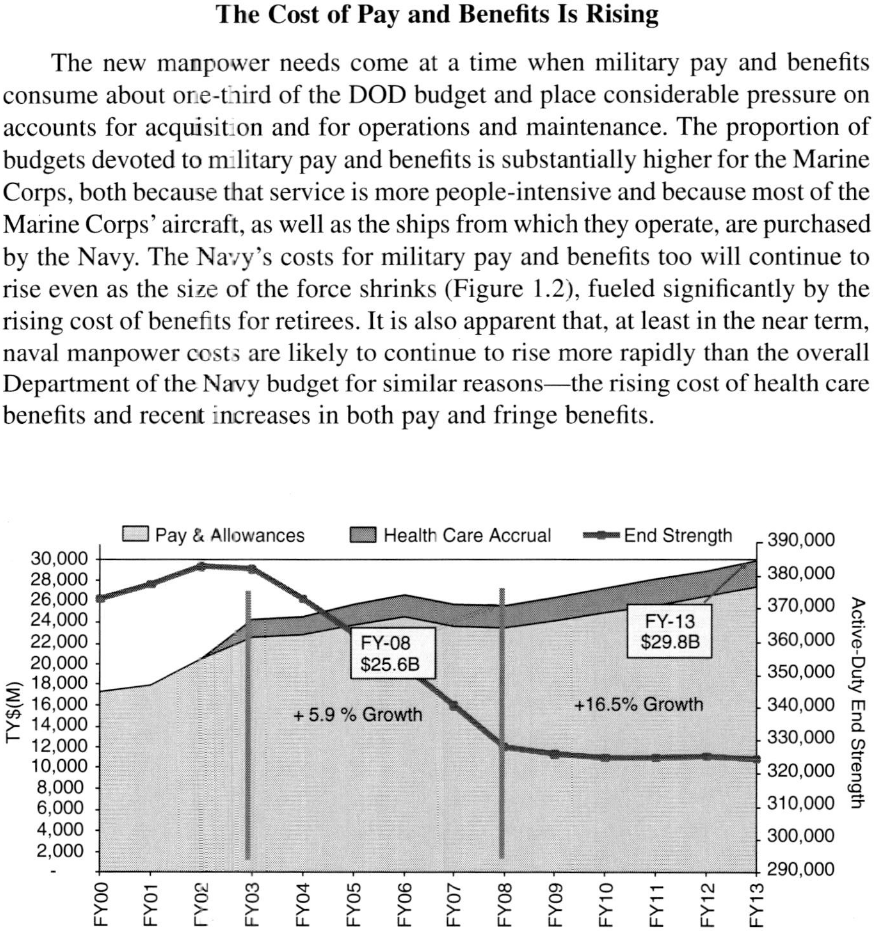

FIGURE 1.2 Changes in active-duty pay and allowances, health care accrual, and end strength with time indicate that Navy manpower costs are rising even as end strength falls. Cost figures are in current dollars, termed then-year (TY) dollars within DOD. SOURCE: VADM John C. Harvey Jr., USN, Deputy Chief of Naval Operations for Manpower, Personnel, Training, and Education and Chief of Naval Personnel, "Shaping the Force for the Future," presentation to the committee, October 3, 2006, Washington, D.C.

By most measures the military compensation increases over the past 10 years have been generous, leading to an overall increase in cost per active-duty troop of 92 percent. Basic pay has grown nearly twice as rapidly as consumer inflation. Like most organizations that provide health care benefits, the military has experienced rapidly increasing costs in those areas especially. Moreover, the costs of retiree medical benefits have grown even faster in recent years than those in the private sector because of major expansions in benefits and the shifting of costs from the private sector to the government.

For example, Tricare for Life, granted by Congress through the National Defense Authorization Act of 2001, substantially increased the health care benefit for military retirees who have reached the age of 65 and are eligible for Medicare. The Tricare benefit for retirees who have not yet reached Medicare age is also growing rapidly in cost. Its cost-sharing arrangement, which has not been adjusted since the mid-1990s, has become so much more generous relative to the beneficiaries' other insurance choices that many in the retired population are shifting the costs of their health care from their postmilitary civilian employers to the government.

Beyond Tricare for Life, Congress in recent years has granted several costly new benefits to the military, even though they were not requested or overtly supported by the DOD or the military services (Table 1.2). The added benefits in fact went beyond the need and contributed to the Department of the Navy decision to request the voluntary separation incentive in order to induce some midcareer personnel to leave the service.

TABLE 1.2 Benefits for Service Members, Families, and Retirees Added Since 1999

New Benefit	History of Adoption
Pay raises above the employment cost index	Requested by chiefs of staff in autumn 1998 testimony; required by Congress for 2000 to 2006
Repeal of Military Retirement Reform Act (1986) retirement provisions	Requested by chiefs of staff in autumn 1998, granted by Congress in National Defense Authorization Act of 2000
Tricare for Life	Granted by Congress in National Defense Authorization Act of 2001, DOD role unclear
Concurrent receipt of military retired pay and veterans' disability compensation	Granted by Congress in 2004 without DOD request
Repeal of Social Security offset for surviving spouse of deceased military retiree	Granted by Congress without DOD request
Expanded Tricare for reservists	Opposed by DOD, granted by Congress

Previous studies have shown that the most cost-effective ways to improve recruiting are to add recruiters and expand advertising.[13] Enlistment bonuses are cost-effective in channeling recruits to specific occupations, but less so in increasing overall numbers. Recruitment and retention also respond to increases in basic pay. Yet the major beneficiaries of most of the recent changes have been the 15 percent of enlisted members and the 50 percent of officers who will retire after serving for 20 years or more, and the retirees themselves. Projections indicate that the costs of benefits for retirees will continue to climb.[14]

These challenges will place a strain on all of the recruited and retained force. People who volunteer for duty do so for their own mix of reasons. Satisfying career tracks and challenging assignments and experiences can be important factors, but studies show that compensation is a crucial motivator of decisions to join, stay, work hard, and endure difficult assignments.

Basic pay is particularly influential because it makes up the largest share of compensation. Because it applies across the board to all service members, however, it is also very costly and generally not useful as a force management tool. In fact, increasing basic pay more rapidly than civilian wages rise can induce many individuals to remain in service longer than force managers would like. Moreover, basic pay is set by Congress and is uniform across all military departments.

The individual military departments have greater flexibility when it comes to bonuses and special and incentive pays, but those pays together amount to less than 10 percent of all immediate cash pay. The Congress, the DOD, and the White House control the remaining 90 percent. The Congress has exerted substantial control over many of the compensation changes in the past decade and is likely to continue doing so in the future.

In an effort to meet the new mission demands and save money, the services have transferred many tasks from active-duty personnel to civilians. The Marine Corps plans to convert 3,404 military billets to civilian billets through FY 2009 and return the "saved" marines to the operating forces.[15]

---

[13]Congressional Budget Office, 2006, *Recruiting, Retention, and Future Levels of Military Personnel,* Washington, D.C., October; James N. Dertouzos and Steven Garber, 2006, *Human Resource Management and Army Recruiting: Analyses of Policy Options,* RAND Corporation, Santa Monica, Calif.

[14]See Congressional Budget Office, 2007, *The Long-Term Implications of Current Defense Plans, Detailed Update For Fiscal Year 2007,* Washington, D.C., April, text adjacent to Figure 2.3 (see <http://www.cbo.gov/ftpdocs/80xx/doc8018/04-20-Defense.pdf>. Accessed on September 17, 2007.). Although pensions for military personnel are paid directly by the Treasury rather than by the Department of Defense, DOD budgets since 1986 have recognized the future costs of retired pay for service members currently in the force through an accrual account. In addition, since the inception of Tricare for Life, the future costs of that benefit for members currently in the force have been recognized through accrual accounting.

[15]Col Otto J. Rutt, USMC, Senior Marine Advisor to the Defense Human Capital Strategy Program Executive Officer (DHCS PEO), "The 21st Century Marine Corps," presentation to the committee, January 23, 2007, Washington, D.C.

Allowing private-sector firms to compete for work done by the government can save money. A Center for Naval Analyses examination of DOD's 2,131 public-private competitions between 1978 and 1994 found that the competitions saved an average of about 30 percent.[16] Such savings may not be achievable for more than a few years and may not be achievable at all when work is complex or difficult to describe.[17] Moreover, while civilians can be deployed to war zones, they are not subject to many of the controls applied to the active force and they can quit![18] Nevertheless, the committee got the impression in its discussions with the military leadership that in the Pentagon the current thinking is that military personnel and civilian contractors are often viewed as interchangeable.

Work Pace and Unexpected Duties Are Taking a Toll

Morale in the service is always a concern and more so with the war in Iraq stretching out. Indeed, the committee believed it sensed early signs of problems during a brief site visit to a few ships at port in Norfolk, Virginia. Appreciating that the committee's site visit was very limited and took place in a port setting—which is a particularly busy time for sailors, with multiple competing interests onshore—committee members were struck by the level of voiced dissatisfaction with several aspects of Navy service, including long hours, lack of training opportunities, inflexible career paths, and individual augmentee (IA) assignments. The committee was particularly surprised at the level of feeling because several meeting presenters assured the committee that end-strength reductions were accompanied by force-shaping procedures and technologies that eliminated major manning problems. The committee's sense was that the cumulative effect of working long days and performing duties that were unexpected or not part of their job descriptions were taking some toll.

The most pronounced example of an unexpected duty was the growing participation of sailors in infantry and infantry support roles in the wars in Iraq and Afghanistan. The Navy has tasked sea, air, and land teams (SEALs), Seabees, medical personnel, chaplains, and EOD (explosive ordnance disposal) personnel, both in units and as IAs for duty in Afghanistan and Iraq.[19] In February 2007 there

---

[16]R. Derek Trunkey, Robert P. Trost, and Christopher M. Snyder. 1996. *Analysis of the DOD's Commercial Activities Program*, Center for Naval Analyses, Alexandria, Va., December.

[17]Cindy Williams, ed. 2001. "Holding the Line on Infrastructure Spending," *Holding the Line: U.S. Defense Alternatives for the Early 21st Century*, The MIT Press, Cambridge, Mass., Chapter 3, pp. 55-77.

[18]The National Defense Authorization Act for Fiscal Year 2007, Pub. L. No. 109-364, Sec. 552 substantially tightened the rules under which government contractor employees operate when they serve with U.S. military forces in a declared war or a contingency operation, by placing them under the Uniform Code of Military Justice.

[19]CDR Daniel Shaw, USN, Individual Augmentation, U.S. Joint Forces Command, "United States Joint Forces Command: Individual Augmentation Requirements," presentation to the committee, December 13, 2006, Norfolk, Va.

were over 13,000 sailors on the ground in Iraq and Afghanistan—more than were afloat in the Persian Gulf. Other issues that the committee noted were overwork due to people shortages (20-hour days), seniors having to perform the work of subordinates because of undermanning, and inadequate time to complete their own assigned tasks. Sailors also felt that their ability to train lower-ranked individuals to do their jobs well was limited by their own work schedules.[20]

The "In Lieu of" and Individual Augmentation programs, whereby sailors are assigned to fill shortfalls in numbers and skills in Army or Marine Corps ground units, is causing personnel turbulence and, at times, seems to hurt morale while creating shortfalls in traditional billets. The sense the committee gained during the Norfolk ship visits was that people join the Navy in part for the seagoing lifestyle and resent being "dirt soldiers." Because of the large number of sailors deployed on the ground in Iraq and Afghanistan, the morale problem created by these programs is not a trivial issue. For example, 300 sailors trained in electronic warfare (EW) spearhead the anti-improvised explosive device (anti-IED) effort in Iraq. The other services have not kept up their EW capabilities and therefore require assistance.[21]

At a minimum, the Navy should inform recruits of the possibility that they can be assigned duty as individual augmentees. Whether the Navy can do much more to make such duties palatable is unknown. The new Navy Expeditionary Combat Command may help, in that people in it will expect to serve in this way.

Based on the presentations and conversations with several of the sailors with whom it talked, the committee believes that the Navy's overall strategy of more automation and system reliability to reduce manual workloads is being used to rationalize the current short-staffing, despite the fact that the new automation and better equipment are not yet in place. Personnel reductions rationalized by reduced workloads due to information technology (IT) enhancements appear to have been taken on some existing ships prior to the full realization of workload reductions. As manpower costs rise faster than the budget and IT becomes more reliable, versatile, and less expensive, future ships are being designed and legacy ships are being refitted with improved IT suites that will reduce some manpower requirements. It is important, however, that IT-associated manpower savings be realized before staffing reductions are made. More generally, care must be taken not to overestimate the manpower savings or the rapidity with which they will be achieved.

The committee believes that the issues noted above reflect work stress. The problems sensed by the committee were reflected more by enlisted members—particularly the senior enlisted—and by junior rather than senior officers.

---

[20]The committee visited the Norfolk piers on December 13, 2006, and held discussions with randomly chosen Navy crew members aboard the USS *Norfolk* (SSN 714) and the USS *Wasp* (LHD-1).

[21]CDR Daniel Shaw, USN, Individual Augmentation, U.S. Joint Forces Command, "United States Joint Forces Command: Individual Augmentation Requirements," presentation to the committee, December 13, 2006, Norfolk, Va.

Some sailors expressed resentment that the retirement system "locked them in" to a poor career option. They projected a corresponding mind-set to leave the Navy as soon as it becomes financially feasible.

Planned Draw-Downs and New Ship Designs Will Exacerbate the Problems

One partial explanation for the discrepancy in perceptions regarding the adequacy of staffing levels the committee detected between formal presentations and informal discussions is that the briefers were using data from the new Smart Ship construct and therefore assumed that ships would require less work and therefore fewer people than are needed today. Unfortunately if today's sailors already feel overburdened, then tomorrow's Smart Ship manning levels and smaller crews are likely to be hard on the multitasked sailors of the future. A top-down look at plans for bringing future ships and aircraft online as they relate to productivity enhancement, including assessments of skill mix adjustments, out-of-norm operations, and lead times for preparing to use the new capabilities, should be considered. The fact that the manpower draw-down is to continue through 2013 adds urgency to the examination of these future productivity issues.[22]

A pressing reason to examine manpower productivity is the near-term addition of new ship designs and retrofitting plans for the legacy fleet. The littoral combat ship (LCS) prototype is presently in sea testing, with delivery of the first operational ships expected in late 2007. The LCS is designed to support three operational modules: antisubmarine warfare, mine warfare, and antisurface warfare against small missile-armed vessels. Basic manning to operate the ship is 40 sailors, and the ship must have its full complement to get underway. The 40-sailor crews will initially consist of experienced, second-term enlistees, called "hybrid sailors," who are capable of handling all of the duties of multiple specialties. The crew for a module will not exceed 35, and there will be only 75 bunks on the ship. Thus, there will be no room for trainees or redundancy.

In the past the Navy would design a ship, assemble a crew, train to a certain level, put to sea, and then come to grips with things that did not work as planned, often by adding more people. With the planned new ships, the Navy cannot afford the luxury of shakedown at sea. Crews must be qualified upon arrival; the Navy calls them "all-up rounds," a term usually used to describe fully assembled ordnance.

The need for experienced crew members will affect both manpower planning and promotion schemes. Currently LCS crews are being handpicked, but there is concern about where all those experienced sailors will come from in the future,

[22]VADM John C. Harvey Jr., USN, Deputy Chief of Naval Operations for Manpower, Personnel, Training, and Education and Chief of Naval Personnel, "Shaping the Force for the Future," presentation to the committee, October 3, 2006, Washington, D.C.

as the Navy expands the LCS fleet to 55 ships. The concern is compounded by the fact that there is no plan to use civilians on the LCS.

Future ship support plans also have profound implications for Navy manpower and personnel. The planned ships will no longer head to sea as self-sufficient units. Instead new ships will be dependent on significant shore-based logistic and electronic support throughout their missions. This shift of work from sea to shore will complicate career management. It may require new ranks and classifications, expansion in some ranks and contraction of others. The changes are likely to affect enlisted personnel more than officers.

The problem will be compounded by similar manning changes projected for all future classes of ships. A legacy cruiser can have a crew of up to 405, whereas the next generation is to have a crew of about 110. A legacy guided missile destroyer (DDG) can have a manning requirement of up to about 370, whereas the next-generation destroyer (DDX) is projected not to exceed 175. Current projections also call for a crew reduction of 1,100 from the current total of approximately 6,000 for future aircraft carriers. Such a reduction has ramifications both for the ship's crew and the air wing.

There is no denying that the Navy is going to become increasingly dependent on more senior, highly trained people, skilled in multiple areas—just the type also coveted by private industry. "Cherry picking" such a large number of crews in the future seems impossible unless the Navy's overall workforce is significantly more capable than today's. The manpower challenges are heightened by the fact that, for example, more than 50 percent of the legacy DDG ships are projected to be in service until at least 2030.[23] As overall manpower is reduced and crew configurations change, preserving the ability to surge in time of threat will be crucial. Chapter 2 discusses in greater detail the staffing needs for the future Navy and Marine Corps.

Marine Corps Mission and Plans

In past years the Marine Corps has been quick to adapt to changes that improve combat capability. The Corps' establishment of air-ground task forces, its evolution of doctrine for amphibious operations and close air support, and its use of vertical lift and jump jets are all examples of fundamental innovation. The Marine Corps differs from the Navy, however, in the degree to which it expects to hold to tradition even as it transforms to meet future challenges and capitalize on technological opportunity.

Marine Corps Commandant General Conway in his guidance states that "we must remain faithful to our enduring mission—to be where our country needs us—when she needs us, and to prevail over what ever challenges we face." He sees

[23] Ronald O'Rourke. 2005. *Navy DD(X) and CG(X) Programs: Background and Issues for Congress*, Congressional Research Service, Washington, D.C.

marines as "Soldiers of the Sea" and the Marine Corps as an amphibious combat force, operating from a sea base, able to respond rapidly and decisively.[24]

Under current plans the Marine Corps will continue to be a young force, with every marine trained as a rifleman regardless of his or her future military occupational specialty. All marines will have the same beginnings: enlisted personnel will all start with the yellow footprints at marine boot camp; officers will all pass through the basic school. All marines will become qualified with their service rifles and acquire competence in basic infantry skills. Through these "touchstone" activities, the Marine Corps expects to keep one foot rooted in the past, while leaning forward with the other foot as it prepares for the future.

The Commandant has directed that the Marine Corps will "invest in science and technology to provide the 'seed corn' for future capabilities and prevent technological surprise." The Marine Corps, he continues, must look beyond the horizon, posture for the future, and develop operating concepts that meet the needs of combatant commanders.[25]

The President's Budget for Fiscal Year 2008 calls for substantial increases in the size of the Army and Marine Corps. For the Marine Corps, authorized end strength is expected to increase by about 5,000 troops per year through FY 2011, to an eventual end strength of 202,000 troops.

The Marine Corps plans to devote about 75 percent of this increase to the operating forces.[26] Some of these increases will be used to fill gaps in current capabilities; the remainder will go toward new transformational capabilities. Specifically, the Marine Corps will increase staffing for intelligence, reconnaissance, and unmanned aerial vehicle activities; expand civil affairs, information operations, and regional expertise; and increase capacity for communications and coalition liaison.

The Marine Corps is also adapting organizationally to the demands of new operations. During February 2006, the service established a Marine Special Operations Forces and Headquarters, a component of the U.S. Special Operations Command. The new component will be compatible with and deployed by Commander, U.S. Special Operations Command.[27]

The Marine Corps has continued to develop transformational concepts of operations and capabilities. Current plans call for expanding seabasing to get

---

[24]Gen James T. Conway, USMC, Commandant of the Marine Corps. 2006. "Commandant's Planning Guide," Headquarters, U.S. Marine Corps, Washington, D.C., pp. 1, 6.

[25]Gen James T. Conway, USMC, Commandant of the Marine Corps. 2006. "Commandant's Planning Guide," Headquarters, U.S. Marine Corps, Washington, D.C., pp. 8, 9.

[26]Col Otto J. Rutt, USMC, Senior Marine Advisor to the Defense Human Capital Strategy Program Executive Officer (DHCS PEO), "The 21st Century Marine Corps," presentation to the committee, January 23, 2007, Washington D.C.

[27]Col Otto J. Rutt, USMC, Senior Marine Advisor to the Defense Human Capital Strategy Program Executive Officer (DHCS PEO), "The 21st Century Marine Corps," presentation to the committee, January 23, 2007, Washington D.C.

marines to distant fights more rapidly, even when adversaries attempt to deny access. Leaders envision seabasing as "a scalable sea-based 'system of systems' that provides operational access and generates strategic speed."[28]

Another area of innovation for the Marine Corps is distributed operations. Leaders hope that distributed operations will enable even small, physically dispersed units to gain the advantage in time and space by combining information and integrating their actions.[29]

A crucial concept for the Marine Corps in recent years is that of the "strategic corporal." Coined by former Commandant of the Marine Corps General Charles C. Krulak in 1999, the term refers to a corporal who is confident in making "well reasoned and independent decisions under extreme stress."[30] In the Marine Corps of the future, even infantry squad leaders are expected to interact confidently and respectfully with foreign populations, understand native customs, and make rapid decisions that reflect credit on their units, the Marine Corps, and the United States. The expectation that most corporals will be strategic corporals has implications for the quality of enlistees as well as their training and career progression.

A related change for the Marine Corps is the transformation in the growing importance it attaches to understanding foreign cultures. The Marine Corps has expanded its Foreign Area Officer and Regional Area Officer (RAO) programs, increasing both the number of officer billets and the areas covered. There are new billets for active-duty civil affairs planners in all major operating headquarters. In addition, the operating forces now include foreign military training units and active-duty psychological operation (PSYOP) teams. The Marine Corps has established the Center for Advanced Operational Cultural Learning (CAOCL), whose mission is to "ensure marines are equipped with operationally relevant regional, cultural and language knowledge to allow them to plan and operate successfully in the joint and combined environment in any region of the world."[31] This program also requires marines to achieve various levels of foreign language proficiency.

---

[28]Col Otto J. Rutt, USMC, Senior Marine Advisor to the Defense Human Capital Strategy Program Executive Officer (DHCS PEO), "The 21st Century Marine Corps," presentation to the committee, January 23, 2007, Washington D.C.

[29]LtGen James F. Amos, USMC, Deputy Commandant for Combat Development and Integration, Commanding General, Marine Corps Combat Development Command. 2007. "Marine Corps Operations in Complex and Distributed Environments," Headquarters, U.S. Marine Corps, Washington, D.C., January 11.

[30]Gen Charles C. Krulak, USMC, 31st Commandant of the Marine Corps. 1999. "The Strategic Corporal: Leadership in the Three Block War," *Marines Magazine,* January, p. 3.

[31]Col Otto J. Rutt, USMC, Senior Marine Advisor to the Defense Human Capital Strategy Program Executive Officer (DHCS PEO), "The 21st Century Marine Corps," presentation to the committee, January 23, 2007, Washington, D.C.

THE WAY AHEAD

Today the naval forces are faced with significant new challenges. Future personnel will all be required to possess a basic level of IT expertise and skills, as discussed in Chapter 2. Personnel with these marketable skills may demand "unequal incentives" to keep them in the services. Uneven promotion opportunities, lateral transfers, incentives to separate from service before becoming eligible to retire, and other "force shaping tools" will be required to meet these challenges.

There have been numerous studies carried out by expert panels, both within the DOD and by outside experts, that have analyzed all facets of the manpower issues. These studies have generated meaningful recommendations, many reinforcing each other from study to study. Unfortunately, however, they have had, for all practical purposes, little effect. Chapter 3 examines and assesses some of these studies and their recommendations.

Navy and Marine Corps requirements for a transformed force demand new manpower and personnel policies that are more flexible and adaptive. Research tools ranging from surveys and analyses of administrative data to pilot demonstrations and experimentation have always been key tools in formulating policies that support military manpower transformation as discussed in Chapter 4.

Chapter 5 identifies several categories of obstacles to change. It then matches several broad categories of manpower and personnel reform to those categories of obstacles. The result is an illustrative mapping of the transformation landscape for naval manpower and personnel policy. Chapter 5 then ends with findings and recommendations aimed at improving the effectiveness of the Department of the Navy in managing and implementing such changes.

2

People for the Future Naval Forces

Future threats, emerging technologies, and new concepts of operations are all combining to place new demands on Navy and Marine Corps people. At the same time, demographic trends and competition from the private sector will make it more difficult for the Navy and the Marine Corps to obtain the people they need with the manpower and personnel policies now in place. This chapter describes the new demands on the Department of the Navy and the demographic and private sector obstacles to meeting them. It recommends new policies designed to remedy this challenge.

FUTURE THREATS, EMERGING TECHNOLOGIES, AND NEW CONCEPTS OF OPERATION

The future security environment is uncertain, particularly for a great power with the depth and breadth of interests possessed by the United States. Uncertainty itself is of course a challenge, demanding a degree of flexibility not required of most nations, for many of whom specific regional threats are clearer and more predictable. There is more than the usual uncertainty facing the United States today because its biggest real and potential challenges occur at opposite ends of the conflict spectrum. The global war on terror (GWOT) defines one end of the spectrum, and the possible emergence of a more technically advanced peer competitor than the former Soviet Union is at the other end of the spectrum.

Of course, the Navy and Marine Corps may not face the full panoply of threats described here at the same time. Moreover, U.S. political leaders can make choices about where to take actions and how to deal with the threats and opportunities the nation faces. Those strategic choices will ultimately be bounded by their political and fiscal costs.

A particularly dramatic choice would be the adoption of a strategy referred to by its proponents as "restraint."[1] Under such a strategy the United States would shun international military involvement and withdraw from most of its alliance commitments. It would greatly reduce the level of forward military presence. Such a strategy might greatly reduce both the operational tempo and the size of the forces that the United States would need. Even under such a strategy, however, the basic trends described below are likely to persist.

New technology creates both opportunities and dangers in regard to the uncertain threat spectrum. The main opportunity is represented by the increasing interdependence of commercial and military technology, particularly in the areas of computing and signal processing. As more and more military technology becomes software rather than hardware driven, it should be possible to create a more agile and responsive force. At the same time the danger is that a legacy force imbued with the traditional focus on military-specific systems will be slow to exploit this opportunity, whereas future opponents may grasp it as a means of leaping ahead.

The new threats, real and potential, together with new technology, have put the entire U.S. military establishment at a crossroads in terms of concepts of operation. For reasons described more fully below, U.S. forces will have to move away from traditional modes of operation, where military forces are aggregated in large units, massed on the battlefield against a clearly defined opponent, and commanded and controlled by central, rear-area command posts with large staffs and senior leaders. Instead U.S. forces will de driven to more distributed operations, where units of action will be smaller, command relationships looser and less direct, and responsibility for decision making on the battlefield will be delegated to lower and lower echelons. At the same time, more complicated combinations of weapons, sensors, and networks linking them together will be wielded by these smaller, more distributed units of action.

The following discussion looks in more detail at these three elements—threats, technology, and concepts of operations—of the new security environment and how together they will challenge Department of the Navy officials who must implement them.

New Threats

The GWOT involves an adversary that avoids direct confrontations with U.S. military forces. Instead insurgents and terrorists operate in small groups, hidden

---

[1]For a discussion of this and other potential grand strategies, see Barry R. Posen and Andrew L. Ross, 1996/1997, "Competing Visions for U.S. Grand Strategy," *International Security,* Vol. 21, No. 3 (Winter), pp. 5-53.

among civilian populations or commercial shipping at sea. They strike at civilian targets or against the forces that seek to protect them. Their military goal is to inflict casualties and cause pain, not to gain and control territory—the latter an objective to be achieved by political rather than military means.

This threat was prominently demonstrated in the terrorist attacks by al-Qaeda on September 11, 2001, and by similar less lethal attacks in Bali, Madrid, and London. The major operations launched by the United States since 9/11, first in Afghanistan and then in Iraq, have exposed U.S. forces and civilian populations in those countries to a lethal environment in which terrorism, insurgency, and civil war blend across neighborhoods throughout entire countries. At the same time special operations forces around the world prosecute a lower-profile, even more dispersed war against individual terrorist cells wherever they are detected and identified. At sea, where terrorist groups and their state supporters must rely on shipping for many of their logistical needs, U.S. naval forces, along with many coalition partners, have instituted maritime security patrols designed to interdict such use of the sea.[2]

At the opposite end of the conflict spectrum, if the United States becomes engaged in a more traditional military combat with a peer or near-peer competitor, it will likely face an adversary that is more integrated with the global economy than was the Soviet Union, and therefore better able to integrate modern information technology into its military forces. In addition, such a competition will occur in the age of persistent wide-area surveillance and precision weapon technology introduced by the U.S. military after the end of the Cold War. So far this has been a one-sided revolution, deployed by the United States against much weaker opponents, but future adversaries will likely deploy similar technology against U.S. forces.[3]

The high end of the conflict spectrum will therefore present two challenges to U.S. forces: the need to stay abreast of a rapidly changing technological battlespace and the need for U.S. forces to present less of a target to their opponents, much as our current opponents do to us today. As will be shown below, these challenges have much in common with those presented at the low end of the conflict spectrum, both with regard to technology and to future concepts of operation.

New Technology

Since the Persian Gulf War of 1991, much has been made about military-technical revolutions and military transformation. These concepts, when applied with any sort of analytic rigor, focus on the ability of new weapons, sensors, and

[2]See, for example, James Bennet with Joel Greenberg, 2002, "Israel Seizes Ship It Says Was Arming Palestinians," *New York Times*, January 5.

[3]National Research Council. 2006. *C4ISR for Future Naval Strike Groups*, The National Academies Press, Washington, D.C.

networks to work together to strike targets with greater precision. The new concepts allow us to detect, identify, and precisely locate targets that were hitherto undetectable or untargetable. The intent is to accomplish this with smaller, more agile, and more dispersed forces linked together to form a network, rather than with a larger concentrated force.[4]

If one looks at the specific instruments of this revolution, one can see that a common, underlying technical theme is the change in the relative importance of hardware versus software, and an ensuing shift in the locus of the cutting edge of militarily relevant technology development from the military to the commercial sector. For example, sensors like radars and sonars will always have an analog interface with the environment they are sensing, in the form of an aperture like an antenna or a sonar dome. Immediately behind that aperture, the electromagnetic or mechanical signals that are collected are now transformed immediately into the ones and zeroes of digital coding and processed using software. Digital signal processing exploits the prodigious advances in miniaturized computational capacity created by the microelectronics revolution, and gives such sensors an unprecedented capacity to distinguish signals from background noise and to exploit those signals to detect, identify, locate, and track targets.

Digital signal processing also enables the creation of reliable, wide-bandwidth communication links that enable networks of sensors, often using different phenomenologies and operating on separate platforms, to work together. Such networks can be put to a variety of new uses, such as the detection and precise tracking of mobile or moving targets or the precise location of radio frequency emitters, such as opposing radars or communication systems. Increasingly it is possible to accomplish these tasks night or day, in all weather conditions, and in real or near-real time.

Accompanying the great advantages created by the ubiquity of software-enabled signal processing are some significant challenges. The U.S. defense industry is dominated by processes and timelines derived from the demands of developing increasingly exotic, military-specific hardware according to strict military-specific standards. Within that context software development focuses on large-scale proprietary solutions that can take years to reach the field. This vitiates one of the prime advantages of shifting to software-driven rather than hardware-driven systems: the speed and flexibility with which a software-driven system can be developed and modified by end users in response to rapidly changing needs.

Certainly the Navy will long face a legacy of large-scale, complicated proprietary software that is difficult and expensive to modify. But even in these cases it is often possible to fashion digital "work-arounds." For example, the transformation of the F-14 Tomcat into a highly capable strike aircraft late in its career was accomplished quickly and cheaply, and without major upgrades to its decades-old

[4]Naval Studies Board, National Research Council. 2000. *Network-Centric Naval Forces: A Transition Strategy for Enhancing Operational Capabilities,* National Academy Press, Washington, D.C.

core avionics, largely with the development of small software interfaces between the legacy systems and the new capabilities being added. Many of these software interfaces were developed by small teams working within the Naval Air Systems Command (NAVAIR). A similar story can be told about the daily challenge of making different communications systems interoperable—a task that can increasingly be accomplished by small teams using tools developed in the commercial software industry.[5]

Of course, to fully exploit the benefits of software-driven systems, they must be designed that way from the start. The first example of this happening on a large scale came when the Navy's submarine community decided to take this approach in the early 1990s with the integrated combat systems aboard its attack submarines. At a time when the antisubmarine warfare (ASW) environment had to undergo radical shifts from deep water to shallow water, and from countering a large, concentrated fleet of Soviet nuclear submarines to a more disparate, global collection of increasingly quiet diesel submarines, it found itself hamstrung in its efforts to update the passive acoustic signal processing routines that formed the core of its sonar capabilities. Out of this challenge emerged the acoustic rapid commercial off-the-shelf (COTS) insertion (ARCI) program. ARCI exploited the fact that its digital combat systems could be converted to a nonproprietary, or open source, software system, which could be managed by a large, decentralized network of trusted providers in both industry and in the laboratories. This allows for a more rapid and continuing means of updating the attack submarine's sonar capabilities. Many groups now compete to provide the latest updates to this system, and the submarine force benefits from the increased flexibility and agility that results.

ARCI still benefits from the fact that it is a large-scale system residing on a relatively small number of platforms and is focused on a threat that may change yearly. Imagine instead the challenge of a system that must reside on hundreds if not thousands of platforms and must be responsive to a threat that changes on a monthly, weekly, or even daily basis. This is the challenge posed to U.S. forces in Iraq and Afghanistan by the wide use of improvised explosive devices (IEDs).

For the purposes of this discussion, an IED is a roadside explosive device that is usually triggered by an operator who remains within the line of sight of the device waiting for U.S. military vehicles to pass. Initially copper cables were the trigger of choice, but the insurgents rapidly transitioned to commercial radio frequency (RF) consumer devices such as garage door openers, walkie-talkies, and cell phones. The counter-IED effort is driven by attempts to electronically attack the triggering device, either by jamming it or, more ambitiously, by predetonating it. This in turn has led to a measure-countermeasure race in which the

[5]On the latter point, see the discussion of the Joint Forces Integration and Interoperability Team (JFIT) in Mladen Rudman, 2007, "Eglin Unit Working to Improve Coordination Between Branches," *Northwest Florida Daily News,* August 11.

United States constantly monitors the environment for new triggers. When one is found the United States marshals a network of electrical and software engineers to counter it, with new solutions arriving electronically within a very short time and quickly distributed via the same means to the many platforms that must use it.

One defining characteristic of the IED threat is the ease with which the opponent can use commercial technology to constantly shift the nature of the threat. A second characteristic is the software-intensive nature of the ongoing response to this changing threat. A third characteristic that is not present in Iraq or Afghanistan, but would be present at the high end of the conflict spectrum, is an opponent able to attack U.S. systems with the same skill sets that the United States is using to counter IEDs. For 15 years the United States has been fighting adversaries lacking the ability to engage in active electronic warfare of their own. It has learned to take largely for granted the ability of its sensors and communication links to operate without interference. This virtual sanctuary would not exist against a peer competitor that was a full participant in the digital electronic revolution. The absence of this sanctuary will drive a far more intensive and extensive software-driven effort not only to attack our opponent's sensors and networks but also protect our own.

The revolution in information technology (IT) is commonly taken as the initiating force behind the acceleration in productivity seen since 1995. The IT revolution has not simply enhanced assembly-line production but has also opened the possibility of fundamentally altering the way production takes place. Productivity, as measured by the amount of work performed within a given timeframe, is often equated with cost: if you could deliver a job at half the cost, many would conclude that your organization had doubled its productivity.

Because of these technical trends, the Navy and Marine Corps will need officers and enlisted personnel who possess high-level IT skills. Such skilled individuals will provide the Department of the Navy with educated buyers and allow units to modify their systems at the pace the threat requires, rather than the pace the defense industry normally provides. To obtain and sustain individuals who are able to ensure such capability, the naval services must develop incentives that recruit and encourage people to be more creative, adaptable, and willing to change. The Navy and Marine Corps will have to make the most of productivity gains promised by IT and modern work practices, satisfy the expectations of workers who have prepared themselves to work in such environments, and create competitive incentives to attract the skilled workforce they require. The committee therefore recommends that the Chief of Naval Operations and the Commandant of the Marine Corps develop a plan for lateral-entry programs to permit and encourage routine entry from the civilian world onto active duty at all ranks for individuals with needed skills. Such programs would also support the global war on terror, where new requirements can suddenly emerge, such as specific cultural and linguistic knowledge for which there is no time to grow the expertise from within.

New Concepts of Operation

If there is one concept that captures the essence of the new concepts of operation that have become necessary in the global war on terror, and will become necessary against a peer competitor, it is distributed operations. Distributed operations entail disaggregating combat forces into smaller, more dispersed operating units.

The global war on terror has driven the U.S. military—and particularly the Army and the Marine Corps—to distributed operations because the threat is distributed. Insurgent groups and terrorist cells operate in small groups, distributed widely among the civilian populations, which provide their cover and also often constitute their targets. These small groups are very difficult to detect and identify before they strike. Therefore, the primary opportunity to defeat them occurs as they strike or in the immediate aftermath. If U.S. forces are to protect civilian populations under attack, they must themselves spread out into small distributed units in order to ensure a rapid response when insurgent or terrorist strikes occur.

The combat that results under these circumstances occurs at the section or platoon level, with relatively junior noncommissioned officers (NCOs) and commissioned platoon leaders in command. Objectives unique to this war place particular challenges on the shoulders of these young marines and soldiers. First, there is a great emphasis on minimizing collateral damage to the civilian populations and infrastructure that make up the battlefield. Second, there is also an emphasis on force protection in order to minimize the casualties to U.S. forces.

These two factors often work at direct cross-purposes to each other at the tactical level. Yet each has an independent and important value at the strategic level. It is difficult under any circumstances to operate in such a way as to maximize discrimination and precision. It is even harder when one must at the same time minimize the exposure of dismounted marines and soldiers, as well as lightly armored vehicles. Nevertheless, it is crucial at the strategic level that both collateral damage and U.S. losses be minimized, because the defense of both the local and U.S. populations is critical in the GWOT.

In practice small units of young marines and soldiers and their leaders must deal with these often competing imperatives by integrating the operational and tactical intelligence with the full range of direct and indirect precision fires available to them. In short, as with many difficult tactical and operational situations in the past, the global war on terror demands a sophisticated combined arms approach. The difference is that the leaders who must accomplish this task will more often be at the platoon or company level, rather than at the brigade, division, or higher echelons.

At the opposite end of the combat spectrum, if U.S. forces find themselves in a military competition or actual combat with a peer competitor, there are equally powerful forces that will drive them toward distributed operations. Here the most important locus of combat or prospective combat will be at sea, and it will be naval forces that must develop distributed operations.

Naval forces already operate in distributed fashion, albeit in a relatively benign environment where the dominant missions are providing air support for marines and soldiers ashore and conducting maritime patrol and intercept operations at sea. Under circumstances in which U.S. Navy forces must project power ashore against an opponent with integrated shore-based defenses and an ability to project defensive power out to sea, the traditional naval response has been to concentrate forces into large battle groups. Concentrated forces simultaneously maximize the collective self-defense capabilities of defensive assets, as well as the efficiency of offensive power projection assets. During the Cold War, this strategy relied upon both organic and off-board sensors that gave the battle group plenty of warning of the approach of air, surface, and subsurface threats. The battle group could reach out over long distances and strike opposing weapon platforms before they could launch their weapons—or in less formal terms, shoot archers not arrows.

Today the prospect for peer-to-peer naval competition is most likely in the Indo-Pacific littoral, against adversaries seeking to deny U.S. naval forces the use of those seas, rather than control them themselves. Such adversaries will likely wield a formidable array of antiship weapons. The most effective platforms for such weapons will be those that are best able to exploit the shallow, cluttered waters that define the littoral environment.[6]

Modern diesel submarines are very quiet, even when snorkeling. When they remain within coastal waters no deeper than several hundred feet, what little acoustic signature they have is dampened both by the short propagation ranges characteristic of shallow water and masked by the noise caused by commercial shipping. At the same time, shallow coastal water poses significant problems for active sonar, both because the bottom is often strewn with shipwrecks and other refuse that could be bottomed submarines, and because of the reverberation that results when powerful sonar signals interact repeatedly with the bottom.

For these and other reasons, detection ranges—and therefore surveillance capabilities in littoral waters—are several orders of magnitude less than they were against Soviet nuclear submarines. Therefore, a strategy designed to ensure that archers are shot before they can loose their arrows is much more difficult to execute. For similar reasons, mines and small missile-armed boats are also a threat. In the former case detection is a challenge, and in the latter case long-range identification is difficult.

Lacking long-range sensors able to detect, identify, and track such weapons, the U.S. Navy realizes that it must deploy a distributed force in order to control the

[6]On these points, see Owen R. Cote Jr., 2006, "The Future Security Environment" and "Sea Shield Past, Present, and Future," *The Future of Naval Aviation,* M.I.T. Security Studies Program, Cambridge, Mass., February, pp. 13-18, 34-38, respectively.

sea space it requires to operate. A distributed force can gain back some of the lost detection range by placing many shorter-ranged sensors and weapon platforms in the contested sea space. For these missions one of the key platforms will be the littoral combat ship (LCS). That means designing the ship to be a "truck" with mobility, damage control, limited self-defense, accommodations, and ability to host a variety of "mission packages."

The underlying LCS rationale demands a large number of ships. The Navy also seeks a platform that can be bought at reasonable cost. A major element of controlling LCS costs will be life-cycle costs, and the main element of life-cycle costs are personnel costs. Hence the Navy's decision to limit the LCS's basic crew to 40, and to limit the crew associated with its separate mission modules to an additional 35.

This is a near order-of-magnitude reduction in crew size compared with similar sized Cold War vessels such as the Oliver Hazard Perry-class frigate. Technology (or the shift of crew functions ashore) will be exploited as much as possible to take up or eliminate tasks conducted by crew members in earlier ship classes, but much obviously remains for the new ships' crews.

The situation with the DDG 1000 is even more striking. Among other things, the Navy plans to expand further on the already formidable capacity and flexibility of the vertical launch missile systems, allowing weapons loads to be tailored to mission requirements and positioning the vertical launch system (VLS) installation itself to contribute to the survivability of the ship.

The new ship designs will translate into big changes on the demand side of the manpower equation. As the new designs increasingly substitute technology for labor, there will be a need for crew members in precise numbers with specific, hitherto unknown combinations of skills. On the one hand, future ships will require a few sailors in whom several sets of skills are fused; on the other hand, these sailors must *always* be on station in the precise numbers needed. There will not be room for error in either the training or distribution system for people at sea.

The greater demands on the crews of both new ship classes will necessitate more collaboration. Smaller ships with more technology to support human performance and improve operational capabilities will require new skills of crew members and arguably will raise the bar for the qualifications of people to learn and apply these skills. Average experience levels will need to increase, leaving relatively fewer jobs for very junior officers and enlisted personnel.

Other programs also impose new demands that the Navy will have to face soon: supplying people with new skills to play key roles in expanded capabilities for the remainder of the improved SSN-688 class and also the converted Ohio-class SSGN with its precision conventional strike capability. The next aircraft carrier will embody changes that will shift the work of some crew members to technology. Future amphibious ships will also trim their crews and demand skills packaged differently in people.

As Figure 2.1 shows for the officer billets, the Navy's rank structure will shift from a pyramidal to an oval structure. Today the largest portion of a ship's crew, whether enlisted or commissioned, are the most junior ranks—the seamen and the ensigns. After the required training and experience at sea, a subset of these enlisted personnel and ensigns are promoted to the next higher rank where there is the need for fewer people. Thus, the current overall manpower structure is a pyramid with the large numbers at the entry levels declining over time to one CNO and one Master Chief Petty Officer of the Navy. In the future there will be fewer jobs in the ship's crew at the entry level because so much more will be demanded of each individual crew member. Hence, the need for the middle ranks, second- and third-term enlisted and commissioned individuals, will exceed the need for lower or higher ranks. The overall force structure will be oval—bulging at the middle ranks—and not pyramidal.

Here, in a very different context, is the same trend toward smaller units of action being asked to take on more and more responsibility. Forces afloat face many of the same demands as those ashore in the GWOT. In this case the trend toward distributed operations is also threat driven, while the trend toward smaller crews is driven in large part by cost, but the result is the same. Regardless of the opponent, both the Navy and the Marine Corps will need people with a more sophisticated set of skills than in the past, and these people will need to be enabled to make battlefield decisions that would have been the province of higher-ranking officers in the past.

Given these demands, the committee offers three findings and recommendations. (The recommendations in the report are not prioritized.)

Finding: Officer career paths in the Navy and Marine Corps are designed to produce well-rounded officers suited for major command and flag rank. However, relatively few officers achieve these goals; most of the career force leave at (or shortly after) 20 years—many at the peak of their operational experience and expertise.

Recommendation: The Secretary of the Navy (SECNAV), the Chief of Naval Operations (CNO), and the Commandant of the Marine Corps (CMC) should take steps to allow those officers who prefer an operational or specialist career to remain in operational or specialist billets, with limitations on rank, for their entire careers. Toward that end the Navy Department leaders should investigate how to embed performance incentives to accommodate career paths that would not involve moving up the chain of command. In addition, the SECNAV, CNO, and CMC should sponsor and support changes to the basic pay table that would allow individuals to stay in grade longer without financial penalty.

Finding: Current promotion patterns from enlisted to commissioned officer are wasteful of talent and inconsistent with the Navy's desire to move toward an "oval

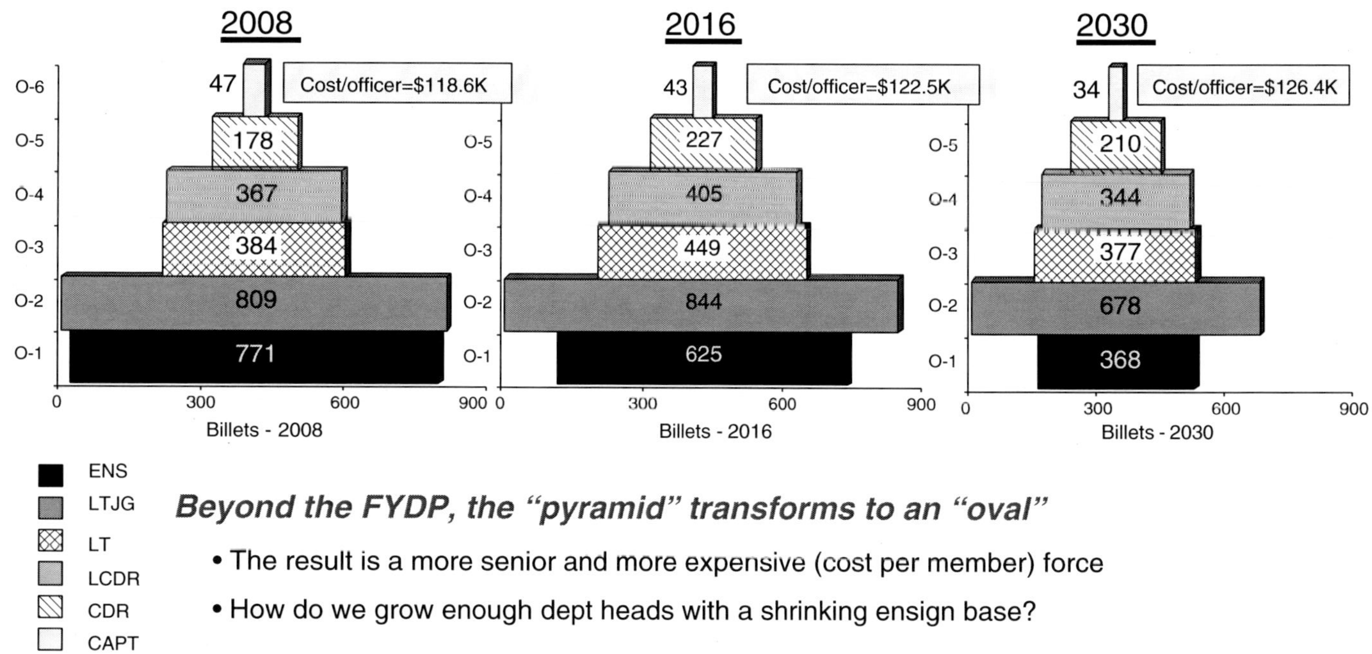

FIGURE 2.1 Surface URL billets on all ships (ship's company). SOURCE: VADM J.C. Harvey Jr., USN, Deputy Chief of Naval Operations for Manpower, Personnel, Training, and Education and Chief of Naval Personnel, "Shaping the Force for the Future," presentation to the committee, October 3, 2006, Washington, D.C. NOTE: URL, unrestricted line (naval officers); FYDP, Future Years Defense Plan.

workforce." The Navy recruits more ensigns than it needs in order to ensure that it will have enough midlevel officers. The surplus of ensigns is inconsistent with the Navy's desire to develop for its future ships a workforce that looks less like a pyramid and more like an oval. At the same time, the Navy provides limited advancement opportunities for E-8s and E-9s, who therefore leave the service in large numbers as soon as or soon after they reach retirement age.

Recommendation: The Chief of Naval Operations should work with the Secretary of the Navy to institute an enlisted-to-commissioned promotion path that pulls senior enlisted people into the midlevel officer ranks, and should use this program to avoid recruiting more ensigns than needed.

Finding: Today's restrictions on entry levels for people will hamper Navy and Marine Corps efforts to exploit emerging technologies and address unexpected threats.

To exploit fully the trend from hardware- to software-driven systems, the Navy and Marine Corps will need officers and enlisted personnel who possess information technology skills at a very high level. Such skilled individuals will have to provide the Department of the Navy with educated buyers and allow units to modify their systems at the pace the threat requires, rather than the pace the defense industry normally provides. These skills are increasingly in demand in the private sector as well.

When a new requirement suddenly emerges, such as the cultural and linguistic knowledge demanded by the global war on terror, there is no time to develop the expertise from within. Whether or not such expertise will be needed over the long term, the short-term need is often urgent. Individuals with such skills are available in the civilian world. If provided the right incentives, such individuals might volunteer to serve on active duty for a specified period. Such individuals would be more valuable to the Department of the Navy than civilian contractors, and such service might be more attractive for the individuals as well.

Recommendation: The Chief of Naval Operations and the Commandant of the Marine Corps should direct the development of a plan for lateral-entry programs to permit and encourage routine entry from the civilian world into active duty at all ranks for individuals with needed skills.

Unfortunately as these threat-driven demands for new skills and additional responsibility are developing, trends in the supply side of the manpower and personnel chain are combining to make it more not less difficult for the Department of the Navy to meet them. In the next section, trends in demography and in commercial labor markets are discussed that will complicate the department's search for the 21st-century sailor and marine to develop into enlightened mature leaders and into operationally savvy personnel such as information warriors, foreign area specialists, and systems engineers.

Many factors affect the ability of the Department of the Navy to acquire the officer and enlisted personnel it needs. In addition to demographics, they include factors such as compensation, the number and motivation of recruiters, mass-media advertising, educational benefits, and bonuses.[7] The unemployment rate and competition from civilian employers are also strong drivers. The next two sections focus on demographics and the competition from civilian labor markets for two reasons: these are the factors over which the Department of the Navy has the least control, and trends in these areas have the potential to undermine existing approaches to recruiting.

DEMOGRAPHICS

As the nation proceeds through the first decade of the millennium it faces a significant cultural diversity shift. The face of America is changing from that of its European founders and settlers to one that is a true melting pot. The melting pot requires new approaches to recruiting and retaining service members. Between 1950 and 2000 the population grew from 151 million to 275 million, an increase of 124 million, or approximately 90 percent. From 2000 to 2050 the population is projected to increase by 144.8 million. In spite of this population increase, the demographic changes (discussed in the following subsection), combined with a lower propensity to serve in the military among young people, will bring new recruitment and retention challenges to the Department of Defense (DOD). The Department of the Navy must be prepared to respond to these challenges.

Demographic Changes

An increasing portion of the U.S. population will be non-white and non-male. The percentage of non-whites in the workforce is steadily increasing, while the white percentage of the total population is expected to decline to about one-half of the population by 2050. Within 50 years Hispanics (an ethnic group, not a race) will almost double their representation, to about one-quarter of the population; Asian representation will more than double, from a bit under 4 percent to about 8 percent; and the growth in the black share of the population will be very flat, going from about 13 percent to about 15 percent in 50 years.[8] (See Box B.1 in Appendix B for definitions of population groups taken from the U.S. Census website.)

---

[7]Barbara A. Bicksler and Lisa G. Nolan. 2006. *Recruiting an All-Volunteer Force: The Need for Sustained Investment in Recruiting Resources*, Vol. 1., No. 1, Strategic Analysis (Policy Perspectives), Arlington, Va., September, pp. 1-25.

[8]The detailed statistics will be somewhat different for the population of 17- to 50-year-olds that includes most service members.

TABLE 2.1 Recruiting Workforce Demographics (Millions)

	2000	2010	2020
Workforce	142.6	155.7	168.5
Males, ages 18 to 24	10.1	10.4	10.5
Females, ages 18 to 24	9.1	9.3	9.9

SOURCE: 2000 U.S. Census (www.census.gov).

Since 1970, the percentage of working-age women who are in the workforce has steadily increased from 43 percent to approximately 61 percent. During the same time period, the percentage of women in the military has increased from 1.4 percent to 15 percent of the force. According to Richard Judy and Carol D'Amico in *Workforce 2020*,[9] in 1997 females received 55 percent of bachelor's degrees, 53 percent of master's degree and 40 percent of doctorates. For the last 15 years black females have attended college and obtained undergraduate and graduate degrees in greater numbers than black males.

Between 2000 and 2020, the size of the prime male recruiting population (ages 18 to 24 years) is expected to increase by only 400,000; the corresponding female population will increase by twice that number (Table 2.1). Unfortunately, DOD-sponsored studies indicate that the military has had limited success in recruiting females in this group. Information from the Youth Attitude Tracking Survey and the Joint Advertising Market Research and Studies provides details on the causes of limited success in recruiting females into the services. The committee raised to the Navy this issue regarding limited success in recruiting females and the Navy's plans on addressing it. The committee heard that at the present time the Navy has no plans that address this issue.

New Workforce Entrants

Workforce 2000 projected that between 1985 and 2000, 85 percent of the new entrants into the workforce would be females and non-whites.[10] Immigrants, both legal and undocumented, represented 22 percent of this total. The statement was widely interpreted as portraying a decrease in the number of white males; in reality the downturn in white male representation was due to rapid growth in the other population groups (see Table B.1 in Appendix B for a more detailed breakdown).

Of these groups the fastest growing is the Hispanic population. As of 2003

[9]Richard W. Judy and Carol D'Amico. 1997. *Workforce 2020: Work and Workers in the 21st Century*, Hudson Institute, Indianapolis, Ind.

[10]William B. Johnston and Arnold E. Packer. 1987. *Workforce 2000: Work and Workers for the Twenty-First Century*, Hudson Institute, Indianapolis, Ind., p. 95.

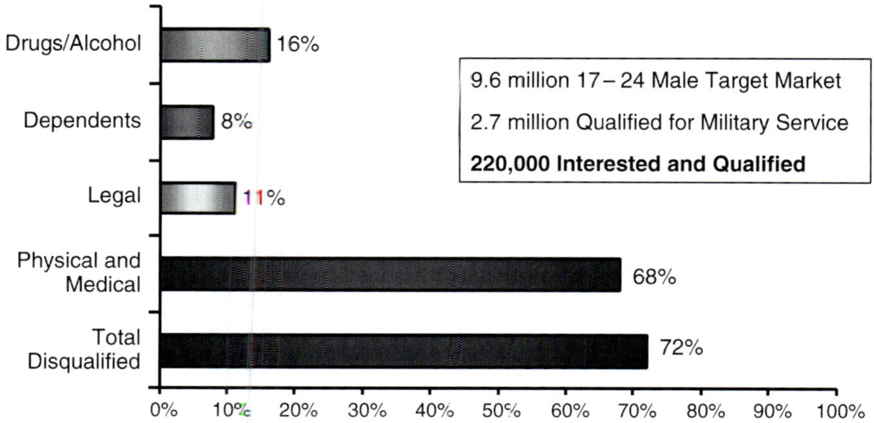

FIGURE 2.2 Non-prior-service recruiting market. SOURCE: Excerpt from RADM Joseph F. Kilkenny, USN, Commander, Navy Recruiting Command, "Navy Recruiting Command . . . Seeking the 'Best and Brightest,'" presentation to the committee, February 13, 2007, Millington, Tenn.

Hispanics replaced blacks as the largest minority group in the U.S. population. In the state of California and a number of major cities nationwide, the majority of the population is now non-white.

Unfortunately for the Navy and Marine Corps, who put great stock in the high school diploma, graduation rates for Hispanics lag significantly behind those of the other groups. With Hispanics now the largest minority group, this issue takes on increasing importance for recruiters (see Table B.2 in Appendix B for a more detailed picture of graduation rates by group).

Propensity to Serve

It was reported to the committee that 72 percent of the 9.6 million male youth in the target age group (17 to 24) are ineligible for military service. Figure 2.2 shows the primary reasons that, of this group, only 2.7 million men were qualified for military service. Allowing for the normal recruiting results, that leaves approximately 220,000 qualified men with a propensity for military service.[11]

[11]It is important to note that not all people who ultimately join the military indicate a propensity to do so when they are surveyed in high school, however. During the mid-1990s, nearly half of all recruits who had been surveyed during high school thought at the time that they would not be likely to join the military. See Bruce R. Orvis, Narayan Sastry, and Laurie L. McDonald, 1996, *Military Recruiting Outlook: Recent Trends in Enlistment Propensity and Conversion of Potential Enlisted Supply*, RAND Corporation, MR-677-A/OSD, Santa Monica, Calif., p. 11.

The DOD sponsors periodic surveys of high school students to gain insight into their likelihood of choosing military services and the reasons they might prefer one service over another. During the 1990s, the Youth Attitude Tracking Study (YATS) found that the propensity to serve in the military was nearly three times higher for men than for women high school students. After the Persian Gulf War began in 1991, the propensity to serve dropped sharply among black men and women and has not recovered since then—and, in fact, has declined sharply since the Iraq War began.

A more recent DOD survey, the June 2005 Joint Advertising, Market Research and Studies (JAMRS) Youth Poll,[12] reported that blacks' propensity to serve was at an all-time low, even though propensity among black females had improved somewhat. Hispanics' propensity was the highest among all youth, but had declined significantly between 2004 and 2005. In the survey the primary reason given by blacks, Hispanics, and females for not wanting to join the military was the ongoing military action overseas. The report identifies positive comments from influencers like family members, teachers, school counselors, and coaches—particularly influencers who have served in the military—as highly instrumental in directing youth toward the military. Unfortunately with fewer and fewer Americans having served in the military, there may be fewer influencers with positive military experiences to pass on to youth. This can only make the task for service recruiters more difficult. Therefore, the committee recommends examining options for expanding programs like the high school Junior Reserve Officer Training Corps (JROTC) and the college Reserve Officer Training Corps (ROTC).

Navy and Marine Corps JROTC was created not as a recruiting tool for the military but rather as a high school citizenship program. It serves to put a uniformed presence on high school campuses and imparts the positive traits associated with the military to this age group. JROTC programs are located in all 50 states, Armed Forces Europe, and Pacific High Schools as well as on Guam. While JROTC is not specifically a recruiting tool, it has been shown over time to have a positive impact on high school students' propensity to serve. Expansion of JROTC would involve establishing relationships between the Department of the Navy and additional individual school districts, since half of the salary of the instructors comes from their military retirement and the other half is paid by the school district.

The Navy college ROTC program is located on 70 college and university campuses. College ROTC is an important recruiting tool, and the variety of locations with participants in ROTC allows the Department of the Navy to target potential recruits with technical expertise (e.g., nuclear, computer technology, information security, engineering, biomedical) and language skills that are critical to a transformed naval force.

---

[12]Joint Advertising, Market Research and Studies, Defense Human Resources Activity. 2006. *June 2005 Youth Poll Overview Report,* Department of Defense, Arlington, Va., p. i.

Role of the Reserve Component

The Reserve Component (RC) is being used at far greater rates today than during the Cold War. Over the last few years the reserves have transitioned from being a strategic reserve, ready to step up in response to a major war, to being an operational reserve, called up regularly to supplement the active component in ongoing operations and deployments. The RC comprises 1.1 million members, or 44 percent of the nation's total military force.[13] Table B.3 in Appendix B shows the changes in RC manning between 1990 and 2005.

While the reservist has willingly entered into his or her reserve obligation, the shift to an operational reserve comes at a price. The cost to the reservist is much more time away from both family and employer, with the return being continued service in defense of the nation. While the Uniformed Services Employment and Reemployment Rights Act (USERRA) offers a degree of employment protection, it cannot make up for not being there. Multiple deployments in rapid succession are having a cumulative negative impact on the structure of the RC and some of its members.

The role of the RC in support of GWOT efforts of the Navy and Marine Corps and other services has been vitally important. The RC provides much-needed force strength and valuable expertise. In addition, the citizen-sailors and citizen-marines provide a special connection to the nation. Following activations or deployments their experiences, both positive and negative, will return with them to their families, communities, and places of employment. The impact of repeated deployments in close succession has the potential of wearing down the support from families and employers that is basic to the viability of the RC.

Assistant Secretary of Defense for Reserve Affairs Thomas Hall has commented that since September 11, 2001, about 352,000 guardsmen and reservists, out of 1.1 million available, have been mobilized (see Box B.2 in Appendix B for a detailed breakdown of various categories of reserves). That number, he said, represents only 38 percent of the drilling reserves, not including individual ready reserve members, while some media reports create the impression that "we've used them all up."[14]

Hall indicated that the fact that 62 percent of National Guard and Reserve members had not been mobilized during the past 10 years had led his department to examine its mobilization patterns and policies. He noted that during the past 6 years, only 4 percent of the force has been affected by multiple mobilizations. But, he added, the department is aware that this is small consolation for those RC members who have deployed more than once. "If you are one of that group that

[13]U.S. Government Accountability Office. 2007. *Military Personnel: Additional Actions Needed to Improve Oversight of Reserve Employment Issues,* Report to Congressional Committees, GAO-07-259, Washington, D.C., February, p. 9.

[14]Kathleen T. Rhem. 2004. "Weekend Warriors No More," *American Forces Press Service, News Articles,* November 8.

has been mobilized, two, three and four times, it's a problem," Hall said. "So we have looked at the stress on the force. We have defined it, and the message here is that it's a small group, but it's an important group."[15]

Hall said that DOD also is looking into providing more predictability for Guard and Reserve members prior to deployment. He said Guard and Reserve members "need to know up front when they're going to mobilize and for how long. We need to get to a predictability, a rotational base for both our active and Reserve forces that provides that."[16]

There has been a continuing partial mobilization of the Reserve and National Guard since the 1990 call-ups following Iraq's invasion of Kuwait. The total number currently on active duty in support of the partial mobilization for the Army National Guard and Army Reserve is 66,912; Navy Reserve, 5,673; Air National Guard and Air Force Reserve, 5,371; Marine Corps Reserve, 5,459; and the Coast Guard Reserve, 303. This brings the total National Guard and Reserve personnel who have been mobilized to 83,718, including both units and individual augmentees (see Box B.3 in Appendix B for details on mobilization categories).

The Government Accountability Office's report *Military Personnel: Additional Actions Needed to Improve Oversight of Reserve Employment Issues* notes that 35 percent of reservists reported working for businesses of fewer than 50 employees.[17] When that 35 percent is combined with the 14 percent who work for firms of 50 to 499 employees, it is the case that almost half of all reservists are employed in what are considered small businesses. These businesses may be the least capable of sustaining the repeated absences caused by activations and mobilizations.[18]

These facts and Secretary Hall's remarks make it clear that the RC today is an *operational reserve* and not a *strategic reserve*. If the changes implied by Secretary Hall are not implemented expeditiously, there will be long-term consequences for the viability of the RC as viewed by employers, coworkers, and their families.

---

[15]Kathleen T. Rhem. 2004. "Weekend Warriors No More," *American Forces Press Service, News Articles,* November 8.

[16]Kathleen T. Rhem. 2004. "Weekend Warriors No More," *American Forces Press Service, News Articles,* November 8.

[17]U.S. Government Accountability Office. 2007. *Military Personnel: Additional Actions Needed to Improve Oversight of Reserve Employment Issues,* Report to Congressional Committees, GAO-07-259, Washington, D.C., February, p. 9.

[18]U.S. Government Accountability Office. 2007. *Military Personnel: Additional Actions Needed to Improve Oversight of Reserve Employment Issues,* Report to Congressional Committees, GAO-07-259, Washington, D.C., February, p. 9. See also the statement of Heidi L.W. Golding, National Security Division, before the Commission on the National Guard and Reserves, May 17, 2005, from the Congressional Budget Office, 2005, *The Effects of Reserve Call-Ups on Civilian Employers,* Washington, D.C., May, p. 2. As the same testimony points out, however, most employers, large or small, are unaffected by the activation of reservists because "only about 6 percent of business establishments employ reservists, and fewer than half a percent of self-employed people are in the reserves."

Finding: Changes in demographics, fitness, and attitudes toward military service call for creative approaches to Navy and Marine Corps recruiting.

Since the inception of the all-volunteer force, the Navy and Marine Corps have devoted substantial resources to recruiting. Those resources have been instrumental in bringing sufficient numbers of high-quality youth into the services over the years and making the all-volunteer force a success.

In the future, military-qualified youth will make up a shrinking share of the U.S. population. Attitudes toward serving in the military may also reduce the pool of potential recruits. At the same time, the nation's racial and ethnic mix is shifting in ways that will advantage employers that emphasize and value diversity and make extra efforts to attract minorities.

Reserve Officer Training Corps (ROTC) scholarships are a crucial tool in attracting qualified students from diverse backgrounds into the officer corps. Complementing that effort, the Junior ROTC offers opportunities for high school students, including many who are otherwise disadvantaged, to improve their physical fitness, build values, and get a taste of military life.

Recommendation: The Assistant Secretary of the Navy for Manpower and Reserve Affairs, the Chief of Naval Personnel, and the Deputy Commandant of the Marine Corps for Manpower and Reserve Affairs should examine options to expand Junior ROTC programs to attract qualified students from diverse backgrounds to naval service.

Congress would have to be persuaded to provide additional funds for Junior ROTC programs. In some cases funds have been appropriated but have not been spent by the services. Location of Junior ROTC units can be critical to the success of the program. School districts with a large minority population need to be emphasized for the new units. Furthermore, there are aviation magnet high schools in the United States that should be made prime targets for units. Under current law only retired active duty members who are drawing retirement can be instructors in Junior ROTC. The way pay is structured, the retired reservist who is not drawing retirement cannot be an instructor. The Secretary of the Navy needs to determine if this requirement can be changed to increase the pool to draw on for instructors.

COMPETITION FROM THE PRIVATE SECTOR

Jobs in all the military services, including the Navy and the Marine Corps, are changing in multiple ways. Many jobs are becoming increasingly complex, and fewer jobs can be performed by individuals with low skill levels. Some jobs require greater technical skills; others require performance of multiple tasks to high standards. Military personnel are increasingly expected to act independently and be adept at thinking on their feet. Because of the rapidly changing environments, equipment, and job requirements, many sailors and marines will be expected to possess the ability to learn rapidly at a high level of mastery.

Increasingly, civilian jobs have the same requirements for quick and adaptive learners who can meet the changing expectations placed on them. Moreover, the demand for highly skilled workers in the private sector will continue to rise.

The empirical evidence of a growing demand for skills shows up in two ways. First, the fastest growing categories of jobs require increasing levels of education. Individuals with only a high school education today are eligible for only 12 percent of new jobs. Second, the large and growing wage premium for workers with higher levels of education reflects unmet demand. Several trends, including technological change, globalization, and demographics, are driving the push for higher levels of skill. Thus, a major source of competition for capable individuals for the military is private industry. Both civilian employers and the military will increasingly compete fiercely to attract, recruit, and retain the highly competent youth.

Every individual makes a career choice. Although some people take the first job offered, most consider multiple factors, including job content, working conditions, pay, and benefits. Some of these factors are generally positive when military service is considered. Others are seen as less favorable when military service is compared with civilian employment. Young people evaluate these factors with varying degrees of accuracy. The evaluation of these factors affects individual job choice and military recruitment and retention. This section of the chapter compares aspects of careers in the military and the private sector and highlights the areas in which the military may be disadvantaged in attracting highly capable youth relative to private industry.

At the conclusion of the chapter the committee offers a recommendation that it believes will enhance the attraction of a Navy or Marine Corps career.

Reputation of the Military Services

Young people who want a career with an organization with a strong positive reputation are likely to find that service in the military, including the Navy and the Marine Corps, meets that need. When asked in a DOD marketing questionnaire about the single most important image associated with the military, 32.1 percent of adults gave responses categorized as "duty/service" and 15.9 percent of adults' responses fell into the "pride/admiration" category.[19] When asked about impressions of those who join a military service, 23.7 percent of adults provided responses in the "heroic" category, 18.4 percent in the "duty/service" category, and

---

[19]Joint Advertising, Market Research and Studies, Defense Human Resources Activity. 2004. *September 2003 Adult Poll 5, Overview Report,* Department of Defense, Arlington, Va., April, pp. 46-47.

13.7 percent in the "pride/admiration" category. Overall, 84 percent mentioned having a positive image of individuals who join the military. In contrast, few U.S. corporations are associated with duty/service, and few corporate employees are perceived to be engaged in heroic activities on the job. A not insignificant number of corporations and their leaders are associated with negative attributes such as greed, lack of respect for employees, and lack of environmental responsibility.

Service to Country

Few employers other than the military offer intrinsic rewards associated with service to country, an important job characteristic for many. While honor and duty to country may be powerful incentives to enlistment for some individuals, they are certainly not incentives for all. For some the desire for extrinsic rewards may override the need for intrinsic ones. And there are other opportunities for service to country (e.g., government work) or service to others (e.g., nonprofit employment). Nevertheless, the opportunity to serve one's country is a powerful attraction for some. It merits noting that the intrinsic rewards may be more important to retention than to recruitment because such rewards may not be apparent until one has served.

Pay and Benefits

Active Component

Many individuals making career choices place a great deal of importance on tangible rewards, including pay, benefits, and other incentives. In both the private and military sectors the value of all forms of compensation is not easy to calculate, making direct comparisons of military and civilian compensation difficult. The military offers basic pay and also adds on housing allowances, uniform allowances, enlistment bonuses, college programs, special-duty pay, and other cash rewards. The private sector may offer base pay, overtime pay, shift differentials, sales incentives, bonuses, overtime, and stock options. In addition, the private sector may offer wage credits for experience or education.

Military pay (including basic pay, allowances for housing and food, and the tax advantage that accrues because the allowances are not taxed) compares quite favorably with pay in the private sector for individuals who have comparable levels of education and years of experience.[20] Among men and women already in uniform, however, misperceptions of the pay package and how well it compares

[20]See Department of Defense, 2002, *Report of the Ninth Quadrennial Review of Military Compensation*, Vols I-V, Washington, D.C., May; and Congressional Budget Office, 2007, *Evaluating Military Compensation*, Washington, D.C., June.

with civilian pay are widespread. For example, focus groups and surveys conducted and analyzed by the Government Accountability Office (GAO) in 2005[21] revealed misperceptions among service members about the structure of their compensation, the costs of their pay and benefits, and the competitiveness of the pay package with compensation in the private sector. Fully 80 percent of service members responding to the GAO survey thought they would earn more as civilians, even though for most of them the opposite was likely true.

Regardless of actual pay differences, perceptions of pay differences may have a strong influence on the likelihood that an individual will enlist or reenlist. When adult influencers believed that civilian jobs paid more than the military, some 60 percent recommended them. The converse held when they believed that military pay was higher.[22]

Reserve Component

Members of the Guard and the Reserve see both pay and benefits advantages and disadvantages. A clear advantage of service is the military pay that supplements pay received from other employment. Additionally, a RAND analysis of Social Security records indicated that the vast majority of reservists earned more when they were mobilized than they did before they were called to active duty.[23] Nevertheless, some individual members suffered a loss in pay when they were mobilized.

Many reservists and their families must change health care plans during deployment, which often means changing health care providers. Others come home to unwelcome surprises regarding health care coverage. For example, one reservist spent 5 months in Iraq in 2003 and was surprised when his employer, a city government, maintained his health insurance but attempted to charge him his employee contributions.[24]

In many respects military health benefits are far superior to those generally

---

[21] U.S. Government Accountability Office. 2005. *Military Personnel: DOD Needs to Improve the Transparency and Reassess the Reasonableness, Appropriateness, Affordability, and Sustainability of Its Military Compensation System,* Report to Congressional Committees, GAO-05-798, Washington, D.C., July, p. 2 and p. 6

[22] See Joint Advertising, Market Research and Studies, Defense Human Resources Activity, 2004, *November 2003 Youth Poll 6, Overview Report,* Department of Defense, Arlington, Va, July, pp. 34; and Joint Advertising, Market Research and Studies, Defense Human Resources Activity, 2004, *May 2004 Influencer Poll 2, Report and Crosstabulations,* Department of Defense, Arlington, Va., November, pp. 3-9.

[23] David S. Loughran, Jacob A. Klerman, and Craig W. Martin. 2006. *Activation and the Earnings of Reservists,* RAND Corporation, TR-274-OSD, Santa Monica, Calif.

[24] See, for example, David Scharfenberg, 2005, "Soldiers Wage Battle at Home Over Wages," *New York Times,* February 27; or the *New York Times* editorial, 2005, "Part-Time Pay for Full-Time Service," *New York Times,* March 10.

offered in the private sector. The number of corporations offering extensive health care coverage to retirees is diminishing every year. Yet these benefits often fail to entice the 17- to 24-year-old group to serve because young people, especially those without families, generally have little need for medical services and fail to see their value.

Available Career Fields

Another factor in many young people's career choice is occupational choice. Some youth have well-developed ideas about what they want to do occupationally. Others clearly do not. Both the military and the civilian sector offer many occupational choices; however, the two sets of occupations are not congruent. Although the array of career fields in the military services in general and in the Navy and Marine Corps in particular is vast, some occupations are excluded. For example, an individual interested in pursuing a commissioned sales career will find limited opportunities. (Many might argue that recruiting duty is a significant exception; however, compensation is not generally commissioned.) The same mismatch, however, is found with civilian jobs as they do not include all the positions one might find in the services. For example, an individual interested in piloting jet fighter aircraft will not find that opportunity in private industry. Although occupational interest is an important determinant of the relative attractiveness of military and civilian careers, civilian occupations are not inherently better or more interesting than military occupations. However, military occupations have higher risk compared to civilian occupations.

For some enlistees the prospect of spending a long career in an occupational field that is not easily transferable to the private sector may be a deterrent to military service. For example, with limited exceptions in law enforcement and private security, expertise in combat arms does not transfer readily to the civilian world.

Not all young people have a well-developed understanding of career options or a desire for a particular one. For some enlistees, career direction from an external source is a positive factor in choosing a career. Few, if any, organizations in the civilian world offer the level of career direction or even have tools equivalent to the Armed Services Vocational Aptitude Battery (ASVAB) to guide new or potential employees. Most seek out individuals whose knowledge, skills, and abilities meet the requirements of the jobs the organization is trying to fill.

Flexibility in Choosing Fields

An advantage of the private sector in the area of career choice for some is the ability to evaluate career decisions and change directions. Increasingly in the United States young people change careers frequently and some continue to experiment throughout their careers. Indeed, the average young person today will change jobs nine times over the course of a career.[25]

In the private sector the growing complexity of business and social environments requires individuals to plan, design, and manage in new ways—taking into account contingencies, anticipating changes, and understanding interdependencies of systems.[26] In doing so individuals are called upon to manage their time and other resources efficiently and effectively in order to execute tasks successfully. With this growing complexity comes the challenge of increasing capacity to do more work while decreasing workforce size and costs.

While the benefits of building tenure and acquiring expertise in a single field often hold employees in a career, other realities like downsizing and offshoring can force individuals in midcareer to make radical changes in career direction. The Department of Labor reported 13,828 "lay-off events" affecting 1,462,069 individuals making initial unemployment compensation claims in 2006. (A layoff event is defined as "fifty or more initial claims for unemployment insurance benefits filed against an establishment during a 5-week period, regardless of duration."[27]) In addition, many midcareer individuals simply grow bored with their chosen careers and strike out on new ones.

In contrast, many military occupations keep a person in a "closed loop" regardless of his or her changing interests. For example, "coners" on submarines typically stay onboard or in support of submarines their entire careers and would have great difficulty making an occupational shift within the Navy.[28]

In many respects officers face the opposite problem. Most are trained for executive leadership even if they prefer to specialize in a career. The opportunity to stay in the military and forego developmental assignments outside a career specialty is limited.

Training Opportunities

Many youth take training and development opportunities into consideration when planning their initial career choices. Both military service and civilian employment offer advantages and disadvantages in this area. Some form of train-

[25]Elaine L. Chao, Secretary of Labor, speech given at the Los Angeles Town Hall, "Challenges to Modernizing the Workforce in the 21st Century" Los Angeles, September 19, 2002. Available at <http://www.dol.gov/_sec/media/speeches/20020919_LA_Town_Hall.htm>. Accessed on October 3, 2007.

[26]See Daniel Goleman, 2000, *Working with Emotional Intelligence,* Bantam Dell Publishing Group, New York; National Research Council, 1999, *Being Fluent with Information Technology,* National Academy Press, Washington, D.C.

[27]Bureau of Labor Statistics, U.S. Department of Labor. 2007. "Mass Layoffs Summary," *News,* USDL 07-1203, August 9.

[28]In the Navy's vernacular, "coners" are the enlisted crew members who serve in the front, or the cone, of the submarine, as opposed to the nuclear engineering specialists who serve in the rear.

ing (e.g., basic training, occupational skills training) is offered to all military recruits.

In contrast, civilian organizations certainly do not all offer formal training to all new hires, and some do not even offer informal training or orientation. New employees learn "OJT" (on-the-job training) or from other employees. Nevertheless, "employers of choice" typically do offer training and offer it in highly desirable career fields. Often the companies developing leading-edge technologies provide the best training in their own technology. As an example, an IT professional may be well advised to seek training from the primary source. Other secondary source training, including military training, can be considered less desirable in terms of both currency and accuracy.

In addition to issues concerning the quality of training, youth may have other significant concerns about training. In some cases particularly in rich labor markets, organizations eschew training altogether in favor of more draconian approaches to acquiring talent. For example, some companies terminate those who lack the right skill set and replace them with new employees whose skills meet the current business needs. Thus, lack of training opportunities is closely related to job security in some segments of the private sector.

A more important training issue than quality of training for many service personnel is the extent to which the training meets the career goals of the sailor or marine. Even high-quality training may not be valued if it is not congruent with the needs of the individual.

Conflict Between Military and Civilian Service

One partial solution to the competition between the military and the private sector for capable individuals is Guard and Reserve service. Both get the services of these individuals, and presumably the individual takes the benefits of both forms of employment. Nevertheless, there are drawbacks.

Civilian employers are required by law to reinstate reservists promptly after they return from active mobilized service. They may not keep employees from military commitments. Yet many organizations find managing their own staffing needs to be a difficult problem that is further complicated by hiring a reservist who may be called to service with little or no notice. Examples of small businesses losing a great deal of money or going bankrupt receive media attention.[29] Small companies are not the only ones to suffer. Large ones also lose valuable skills when reservists are called to serve.[30]

The transition of the reserves from a strategic role to an operational role has sharpened the problem. In a 2004 interview Assistant Secretary of Defense for

[29]Lee Roberts. 2006. "Solutions: A Call to Duty Can Overwhelm Small Companies," *New York Times,* September 12.

[30]Fara Warner. 2003. "Business; Carrying On When Suits Are in Camouflage," *New York Times,* April 13.

Reserve Affairs Thomas Hall pointed out that service in the reserves is no longer simply weekend duty and emphasized that recruiters need to make clear to recruits the demands that will be placed on them for more "robust" service.[31]

Military Lifestyle

Perhaps the most notable difference between military and civilian employment is the military lifestyle. There are advantages and disadvantages to military life, and the value of many aspects depends on the individual and his or her needs and desires.

Many find the demands of the military life daunting. Whereas most civilian employees may work an 8-hour day with occasional overtime, the sailor or marine can be asked to work far beyond 8 hours on a regular basis. Military personnel must adhere to certain physical standards, and their behavior in personal life is often subject to review. In addition to the differences in daily life, absences from home required of Navy and Marine personnel (e.g., sea duty, deployments, and individual augmentee assignments) can be particularly difficult for families and may impede the fulfillment of personal goals, such as education. Despite the seemingly negative aspects of military lifestyles and job requirements in times of war, some individuals thrive on the excitement and danger of deployments to war zones.

Virtually all occupations present some form of stress. Data comparing the stress induced in civilian and military careers or the perceptions of the stress levels are not available. However, there is evidence that the stress during first-term enlistments is high. Using data collected between April 2002 and May 2005, Harris et al. suggested that stress levels are high and concluded that stress during initial training in the Navy plays a role in future attrition.[32]

It is important to note that the military lifestyle has some positive features as well. Perhaps the foremost is the sense of community many find in military service. Marine Corps officers suggest that some young people, particularly young Hispanic males, bond closely to their fellow marines and note that recruiters often approach groups of Hispanic men for enlistment, promising that all of them can go to boot camp at the same time.

All the services, including the Navy and Marines, appear to build formal and informal infrastructures that support sailors, marines, and their families. The Marine Corps particularly capitalizes on its ability to establish community and extends the community in its Marine for Life program, which supports former

---

[31]Kathleen T. Rhem. 2004. "Weekend Warriors No More," *American Forces Press Service, News Articles,* November 8.

[32]Rorie N. Harris, Michael. A. White, Naina C. Eshwar, and Jacqueline A. Mottern. 2006. "Stress and Attrition from Military Training: First-Term Sailors in the U.S. Navy," paper given at the 48th Annual IMTA [International Military Testing Association] Conference, October 3-5, 2006, Kingston, Canada.

marines throughout their lives. The level of support provided employees in corporate America generally falls far short of that provided military personnel.

Standards for Service

Unlike the private sector, the military has extensive standards for education, aptitude, physical and medical condition, moral character, and citizenship or residency status for enlistment.[33] In addition to the DOD minimums each branch may set standards that are more stringent. In contrast, most private enterprises do not place so many restrictions on applicants. While citizenship requirements, review of criminal activities, drug testing, educational standards, and demonstration of abilities are somewhat common, medical and physical standards are not. In fact, civilian employers are effectively barred from setting such standards by the Americans with Disabilities Act, which prohibits inquiries regarding current and past medical problems. The net effect of the military enlistment standards is to further reduce the pool of potential enlistees.

Work and Family Balance

A characteristic of modern American life is the many demands for time that come from work as well as personal life, including family. Many American workers in the private sector struggle to balance the needs of their jobs and those of their personal lives; however, few are expected to be away from home for long periods of time. As noted above, absences from family are required in the military, particularly in a time of war. These absences are assuredly difficult for all service members, but some groups may be more challenged to balance the needs of their families with their military service. Personnel with children may miss parenting opportunities during their children's formative years. Single parents may not find sufficient resources to assist them in meeting the demands of parenthood. Mothers of infants may find their children's needs incompatible with long absences.

Working Conditions and Personal Danger

Jobs in both the civilian and military sectors can entail unpleasant working conditions, though few present as much potential for personal danger as the military. Although *Today's Military*[34] reports that 81 percent of the 4,100 military

[33]For a comprehensive discussion of physical and health standards, see National Research Council, 2006, *Assessing Fitness for Military Enlistment: Physical, Medical, and Mental Health Standards,* The National Academies Press, Washington, D.C., Chapter 2. The report also examines the changing physical and mental condition of U.S. youth and reviews the military usefulness of today's standards.

[34]See "Myths vs. Reality" at <www.todaysmilitary.com/app/tm/parent>, an informational website provided for parents whose children are considering opportunities available through the military. Accessed July 9, 2007.

occupations are noncombat positions, a reality of military service particularly in times of war is the danger. The number of U.S. soldiers and marines killed in Iraq and Afghanistan is reported daily.

While many in the 17- to 24-year-old age group may not recognize or fear these threats, many of their career influencers, particularly their mothers, do. Mothers are one of the demographic groups least likely to recommend military service, with only 37 percent favoring service compared with 42 percent of fathers and 70 percent of former military service members.[35]

What's Different Today?

Many aspects of military life have been in place for years, and the differences between military and civilian employment have been observed over a long period of time. Obeying orders, meeting standards, and accepting the lifestyle have been required for generations, and the services have generally met their recruiting goals. If the services continue to put today's levels of effort and resources into recruiting and retention, they may be rewarded with success. Yet a few things raise concerns about the degree of difficulty they may face.

First, as discussed in the first section of this chapter, changes in equipment and staffing plans in the Navy and Marine Corps will require an increasing number of sailors and marines with highly marketable skills at the same time that the demand for the same skills in the private sector is increasing. Although census statistics indicate slight growth of the youth population, that segment is diminishing relative to other age groups in the population. There may simply not be enough youth who meet the requirements of the military, given the competing demands of the private sector.

Second, it is not clear how these able youth will respond to the "negative" aspects of military life, as these people have many attractive options in the civilian world. In past years youth may have found fewer good employment opportunities in the private sector and may have been more willing to accept military service.

Third, the country is engaged in a long war that highlights the dangers and privations inherent in military service, which are obvious deterrents to military service for many young people as well as many of their influencers.

Fourth, many youth are disenchanted with national politics and as a result with all federal service.

Fortunately there are counterbalancing forces favoring the military. First, the global war on terror initially had a galvanizing effect on patriotism. To the extent that the American public continues to believe that the military is serving a legitimate role fighting terrorism, the military may continue to attract its fair share of capable youth.

---

[35]Joint Advertising, Market Research and Studies, Defense Human Resources Activity. 2004. *September 2003 Adult Poll 5 Overview Report,* Department of Defense, Arlington, Va., p. 35.

Second, recruiting consistently improves during downturns in business cycles of the American economy. Recessions have the immediate effect of diverting capable youth to military service in the absence of good private-sector alternatives, and the longer-term effect of reminding youth that the military may be a more reliable long-term employer. Although the timing of economic cycles is difficult to predict, the likelihood that recessions will follow periods of economic expansion is almost certain.

No single policy change will address all of these challenges. As discussed earlier, many find the demands of military life daunting, and so interludes would be good. For example, the committee believes the provision of career on/off ramps would help sailors and marines take time off from active duty in order to obtain education, take advantage of training opportunities, or start a family without undue prejudice to their careers. This policy would both reduce the early exit of qualified people and potentially increase the attractiveness of a military career for those who are now deterred from pursuing this option. In particular, the provision of career on/off ramps could prove instrumental if the Department of the Navy finds its dependence on female sailors and marines increasing because of the demographic trends described above. In Chapter 4 the committee proposes a controlled randomized experiment for on/off ramps.

Finding: Today's policies can make a return to active-duty service challenging or unattractive for sailors and marines who wish to take time out to obtain training and education not offered through the services, gain experience in the civilian world, or start a family.

Today some sailors and marines of both genders leave for educational or workplace opportunities that are not available to them on active duty. Women early in their careers exit the services in disproportionate numbers. While they may need only a few years off, those who leave generally do not return to service. The early exit of experienced sailors and marines puts an extra burden on the remainder of the force and on resources for recruiting and training. A major challenge with no obvious solutions in any organization, whether in the private sector or the military, is to find ways not only to create effective on/off ramps but also to ensure that the careers of people who do take time off progress commensurate with their skills and experience.

Recommendation: The Chief of Naval Operations and the Commandant of the Marine Corps should direct the development of programs for on/off ramps whereby sailors and marines could without undue prejudice to their careers take time off from their active-duty careers in order to obtain education, take advantage of training opportunities beyond those provided by the Navy and Marine Corps, or start a family.

3

Summary and Assessment of Previous Studies

In considering the manpower needs for a transformed force, it is useful to review past research and study group efforts for insights relevant to the Department of the Navy. The Preface highlights two such efforts: the 2006 Department of Defense (DOD) report of the Defense Advisory Committee on Military Compensation (DACMC)[1] and the 2005 report by the Center for Naval Analyses (CNA), *Military Compensation Reform in the Department of the Navy*.[2,3] The CNA report was itself a compendium and review of recommendations from earlier studies. In addition, the committee reviewed studies of the effects of deployment on retention later in this chapter.[4]

[1]Defense Advisory Committee on Military Compensation. 2006. *The Military Compensation System: Completing the Transition to an All-Volunteer Force,* Arlington, Va., April.

[2]Michael Hansen and Martha Koopman. 2005. *Military Compensation Reform in the Department of the Navy,* Center for Naval Analyses, Alexandria, Va., December.

[3]A third document highlighted in the Preface is the *Department of the Navy Human Capital Strategy,* which is addressed in Chapter 5. See William A. Navas Jr., Assistant Secretary of the Navy, Manpower and Reserve Affairs; LtGen Garry L. Parks, USMC, Deputy Commandant, Manpower and Reserve Affairs; and VADM Gerald L. Hoewing, USN, Chief of Naval Personnel, 2004, *Department of the Navy Human Capital Strategy,* Department of the Navy, Washington, D.C., June 21.

[4]See James Hosek and Mark Totten, 1998, *Does Perstempo Hurt Reenlistment? The Effect of Long or Hostile Perstempo on Reenlistment*, RAND Corporation, MR-990-OSD, Santa Monica, Calif.; James Hosek and Mark Totten, 2002, *Serving Away From Home: How Deployments Influence Reenlistment*, RAND Corporation, MR-1594-OSD, Santa Monica, Calif. A similar result is found for officers; see Ronald Fricker, 2002, *The Effects of PERSTEMPO on Officer Retention in the U.S. Military*, RAND Corporation, MR-1556-OSD, Santa Monica, Calif.; Heidi Golding with David Gregory, 2002, *Sailors' Willingness to Complete Sea Tours: Does Money Matter?*, Center for Naval Analyses, CNADRM 00006886/Final, Alexandria, Va., December.; and James Hosek, Jennifer Kavanagh, and Laura Miller, 2006, *How Deployments Affect Service Members*, RAND Corporation, MG-342-RC, Santa Monica, Calif.

This chapter summarizes those studies and critiques their recommendations based on their potential for enabling the Department of the Navy to meet its future manpower and personnel objectives. The chapter begins with brief summaries of the DACMC report and the CNA study. It continues with an explanation of the criteria that the committee considered in assessing these studies and provides a critique of the findings and recommendations of these studies. The next section summarizes the deployment studies. The chapter concludes with a summary of the committee's findings and recommendations related to the studies it reviewed.

SUMMARY OF DEFENSE ADVISORY COMMITTEE ON MILITARY COMPENSATION REPORT

The DACMC was chartered in March 2005 to advise the Secretary of Defense on military compensation. The committee, led by Admiral Donald Pilling (USN, retired), conducted research and held public meetings over the course of one year, resulting in a final report in April 2006.[5]

Like several other recent studies, the DACMC report explores changes to pay and benefits aimed at ensuring the future recruitment and retention of high-quality personnel in a cost-effective manner to sustain a ready force.[6] The DACMC believed that the Secretary of Defense and services know best how to manage the force. Its intent was not to provide specific recommendations and precise military compensation policy changes. Rather, it provided a framework for guiding change, and a set of recommendations based on that framework. The following paragraphs summarize the guidelines recommended by the DACMC and their study recommendations.

Guidelines for Change

The DACMC's framework calls for changes to the military compensation system that will help maintain the volunteer professional force. Changes should be aimed at improving force staffing, force management, motivation, and per-

[5]Defense Advisory Committee on Military Compensation. 2006. *The Military Compensation System: Completing the Transition to an All-Volunteer Force,* Arlington, Va., April.

[6]For example, the Defense Science Board (DSB) Task Force on Human Resources Strategy in 2000 also addressed this issue (Defense Science Board, 2000, *Human Resources Strategy,* Office of the Under Secretary of Defense for Acquisition, Technology, and Logistics, Washington, D.C.), as did the 2001 Quality of Life (QoL) Panel led by Admiral David Jeremiah, USN, retired (Naval Research Advisory Committee Report, 2001, *Quality of Life: Reviewing Commitment to Our People,* Office of the Assistant Secretary of the Navy for Research, Development and Acquisition, Washington, D.C.). The DACMC, cited in footnote 1, also builds on a longer history of compensation study groups, including nine Quadrennial Review of Military Compensation studies. Because of the long history of compensation studies, including the recent DSB and QoL panel reports, the DACMC was able to draw on a rich and well-developed literature. The same is true for the CNA study cited in footnote 2. The tenth Quadrennial Review of Military Compensation is underway as of this writing.

formance, and at making the system more efficient. That is, the changes should help meet the goals of any compensation system, namely, to attract, retain, and motivate the right quantity and quality of service personnel, assign them where they are needed, and do so cost-effectively. The specific criteria or principles recommended by the DACMC as guidelines for changing the active and reserve compensation system were the following:

1. *Force management*. Changes to compensation should be tied to force management goals.
2. *Flexibility*. The compensation system should adjust quickly to changes in supply and demand of personnel in general and specific skill areas.
3. *Simplification*. Changes to compensation should promote simplicity rather than complexity to improve understanding of the system by members.
4. *Systems approach*. A change should consider the force management implications for active and reserve personnel.
5. *Choice, volunteerism, and market-based compensation*. Changes should promote choice and member preference rather than involuntary assignment.
6. *Efficiency*. Changes to compensation should meet objectives cost-effectively.
7. *Cost transparency and visibility*. The full costs, over time, of changes should be clear to policy makers.
8. *Leverage*. Where possible, compensation improvements should leverage benefits in the civilian sector rather than crowd them out.
9. *Fairness*. Commitments should be honored.

DACMC Findings and Recommendations for Change

Overall, the DACMC concluded that current compensation levels are adequate to meet staffing needs, but the structure of compensation should be improved to give force managers more flexibility and to increase the efficiency and effectiveness of the compensation system. The DACMC focused on six major topic areas.

Retirement Reform

The DACMC argued that the current system for active-duty service members is inequitable, inflexible, and inefficient. The system is inequitable because most members never reach 20 years of service and thus never receive benefits. It is inflexible because its one-size-fits-all approach leads to an overuniformity of career lengths, which in turn impedes flexible force management. Career lengths are too short in many career fields, but too long in others. The system is inefficient because it backloads military compensation in deferred compensation that most members will never receive and do not value highly. The DACMC points out that the typical military member is relatively young, and young people tend to discount deferred benefits at high rates. As a result, it costs more to improve retention by increasing deferred compensation than by increasing current compensation. The military compensation system would be more efficient if a higher share were in current rather than deferred compensation.

The DACMC recommended a three-part reform of the active-duty military retirement system:

1. A 401(k)-like plan, vesting after 10 years of service, to which the government would contribute in the range of 5 percent of basic pay;
2. A defined-benefit plan that would pay an annuity beginning at age 60, vesting after 10 years of service, using a formula similar to the current retirement annuity formula; and
3. Additional current compensation to achieve force management goals. Such compensation might come in the form of separation or transition pay of limited duration for those who leave after the vesting point; increases in basic pay or bonuses; or "gate pay" for members completing key career milestones,[7] such as at 10, 15, 20, 25, and 30 years of service.

Pay for Performance, Pay Table Reform, and Pay Differentials Based on Dependency Status

The DACMC found that current policies reward performance almost exclusively through the promotion system, with the main financial incentive being higher basic pay. The current pay table specifies pay as a function of a member's pay grade and time in service or longevity. Consequently those who are promoted earlier than their peers experience only a short-lived financial advantage. Once peers are promoted, the financial advantage disappears. Thus, the overall pay advantage for those who perform especially well is relatively small.

Another disadvantage of the current time-in-service pay table is that it does not account for civilian experience. Because pay is tied to tenure in the military, the services are constrained in attracting either lateral entrants who developed needed skills in the civilian world or individuals with prior military service who wish to reenter the military after gaining experience on the outside.

The DACMC recommended changing the pay table to make basic pay a function of grade and time in grade, rather than grade and years of service. Those promoted earlier than their peers would enjoy a larger permanent pay advantage. For individuals entering the services from civilian life, the shorter total time in service would not make their pay lower than that of their peers with similar total levels of experience.

---

[7]"Gate pay" is additional pay or a bonus that is a multiple of basic pay and is payable at key years of service such as 10, 15, 20, 25, and 30 years.

The DACMC also examined allowances for housing. Today housing allowances for members with dependents are about 25 percent greater on average than for members at the same grade and year of service who do not have dependents. This difference in compensation is not related to a difference in productivity. The bigger paychecks for those with dependents provide an incentive for members to marry or have children and for members with families to stay in service. The DACMC recommended that this differential be eliminated by paying the housing allowance to every member at the "with dependents" rate.

The DACMC also recommended that all members receive a housing allowance, whether or not they live in government housing, and that those in government housing pay fair market rental rates. This may mean that some members, especially junior enlisted members living in barracks, would receive housing allowances in excess of the new rents they would pay. Finally, the DACMC recommended that the family separation allowance (FSA) be consolidated with other special and incentive pays related to deployment or unaccompanied tours, arguing that the appropriate level of compensation for deployments should not differ between those with and without dependents.

Special and Incentive Pays

The DACMC found that while special and incentive pays are an important part of the overall military compensation package, they are less effective than they could be. First, managing and monitoring the large number of special and incentive pays is burdensome. Second, the payment criteria and amounts for some of the pays are set rigidly in law and difficult to change. Furthermore, some of these amounts have become entitlements, paid without regard to force management considerations. Third, special and incentive pays are the services' main tool for flexible force management, yet they constitute less than 5 percent of immediate cash pay.[8]

The DACMC recommended that special and incentive pays be consolidated into a smaller number of categories. Within each category the budget would be fungible; the Secretary of Defense and secretaries of the military departments would have authority to determine criteria and to set and change payment amounts. The DACMC also recommended that special and incentive pays be increased as a share of compensation. Once increased, the effectiveness of these pays in achieving force staffing goals should be assessed to determine whether the increase in share should be sustained.

[8]See P.F. Hogan, C.J. Simon, and J.T. Warner, 2004, "Sustaining the Force in an Era of Transformation," in *The All-Volunteer Force: Thirty Years of Service,* B.A. Bicksler, C.L. Gilroy, and J.T. Warner (eds.), Brassey's, Inc., Washington, D.C.

Military Health Care Benefit

The DOD offers a health care benefit called Tricare to active-duty military personnel and their families as well as to retirees and their spouses and surviving spouses. For active-duty members and their families, Tricare provides comprehensive health care with virtually no premium or out-of-pocket expenses.

The nature of Tricare's coverage for military retirees and their spouses depends on the individual beneficiary's eligibility for Medicare. If the retiree or spouse is Medicare-eligible (usually that means age 65 or older), then Medicare is the first insurer, and Tricare, which in this case is called Tricare for Life, provides wrap-around coverage. Retirees and spouses who are not yet eligible for Medicare can choose from among several coverage options, much as in a private-sector plan.

The DACMC found that Tricare for Life, the wrap-around coverage offered to military retirees who are eligible for Medicare, has shortcomings. Although it is quite expensive to provide, it is not highly valued by the typical active-duty service member, who is relatively young and is not fully appreciative of benefits that will not accrue until age 65.

The DACMC also found that the premiums and other fees paid by pre-age 65 retirees for their Tricare coverage are substantially lower than those of plans typically offered by civilian employers. As a result Tricare tends to crowd out any civilian health benefits that might be available to this group. Yet policy makers have little incentive to minimize the cost of pre-65 health care, because any savings from policy changes would not be realized until far into the future.

To address these issues the DACMC recommended that premiums and other fees under Tricare Prime, the health maintenance organization option under Tricare, be set for pre-65 retirees to be more consistent with the cost-sharing provisions under typical civilian employer plans. To stay in line with civilian plans those Tricare fees should grow at the same rate as the annual cost-of-living adjustment to the military retirement annuity.

The DACMC also recommended that Tricare for pre-65 retirees be funded on an accrual basis, similar to the current retirement system and to the Tricare for Life program for post-65 retirees. Putting the program for pre-65 retirees on an accrual footing would force policy makers to recognize today the future costs of retiree health benefits for all currently serving members. The move should improve incentives for leaders as they consider policies related to end strength, tenure profiles, and health care programs.

Quality of Life

The DACMC focused on two quality-of-life issues, spouse employment and dependents' schools. It argued that because of frequent moves and assignments to rural areas, military spouses face reduced civilian opportunities relative to civilian spouses. In addition, the educational opportunities for children and spouses are less than desired. Greater choice in assignment, along the lines of what the Navy offers under the assignment incentive pay program, provides opportunities to avoid problems and to reduce the hardships associated with less preferred assignments, including less desirable employment and educational opportunities for spouses and children.

The DACMC pointed out that quality-of-life programs can offset some of the hardships and challenges of military life for members and their families. The DACMC agreed with the economics literature, which finds compensation delivered in-kind to be generally less efficient than cash, because cash can be used flexibly for any purpose whereas noncash compensation cannot. Other things being equal, people tend to prefer cash to a commodity of the same cost.

The DACMC also noted that it is generally difficult to quantify the effect of quality-of-life programs on retention or readiness. Because these programs are typically small in scale, their predicted effects on retention are generally modest. Consequently little information is available on the cost-effectiveness of quality-of-life benefits. The DACMC argued that the ultimate responsibility for quality of life rests with the commanders of military units. Therefore, it is important that commanders of units ensure that members understand and take advantage of available quality-of-life programs.

The DACMC recommended a rigorous periodic reevaluation of these programs, to ensure the best allocation of resources between cash and in-kind benefits. Such evaluations must recognize the difficulty of quantifying the effects of quality-of-life programs on retention and other metrics of readiness. The DACMC therefore recommended that DOD develop guidelines for determining the effectiveness of quality-of-life programs.

Reserve Compensation

The DACMC acknowledged the changing role of the reserves, from the strategic missions of the Cold War to an operational role, under which more reserve members are now being activated and for longer periods. The DACMC noted that while the Army's Reserve Component fell short of recruiting targets in recent years, reserve recruiting and retention in the other services have remained adequate, despite increased usage in support of Operation Enduring Freedom and Operation Iraqi Freedom.

When reservists are mobilized, they and their families become eligible for Tricare. If they were previously covered by the health plan of a civilian employer, switching to Tricare may mean changing doctors and other health care providers. The DACMC found that many reserve members attempt to maintain continuity of care by keeping their civilian employers' health insurance when they are activated.

The DACMC recommended guidelines for setting reserve compensation.

First, it recommended that activated and mobilized reservists should receive the same pay and benefits as their active duty counterparts. Second, it recommended that compensation policy should be coordinated to the extent that active and reserve components recruit from the same markets, as in the case of nonprior-service personnel, and that prior-service reserve personnel are drawn from those who separate from active duty. Third, it recommended that the reserve components be given the flexibility to set recruitment and retention bonuses that vary by geographic location and or unit. Currently such bonuses vary by occupation. To the extent that staffing problems arise at the unit and local level, locale- or unit-based bonuses could help the reserves resolve these staffing issues.

SUMMARY OF CENTER FOR NAVAL ANALYSES REPORT

To identify promising compensation tools to meet the Navy's 2004 human capital strategy (HCS), the Assistant Secretary of the Navy for Manpower and Reserve Affairs (M&RA) and the Deputy Assistant Secretary of the Navy for Total Force Transformation (DASN[TFT]) formed a compensation team in 2005. The team drafted guiding principles and strategic goals and asked the Center for Naval Analyses to assess major compensation tools in light of these goals. In brief, the goals, drawing from the HCS, are to recruit, retain, motivate, and develop personnel who are of high quality in terms of education and aptitude, assign them to needed positions, and ultimately transition them voluntarily to civilian life when it is best for them to leave from the standpoint of the Department of the Navy.[9] The guiding principles were that the compensation tools support the all-volunteer military, promote flexible workforce management, enhance cost-effectiveness, and support the achievement of strategic objectives.

CNA assessed the Navy's compensation tools in terms of these goals and principles based on evidence from the available literature. The compensation tools that CNA assessed were as follows:

- Regular military compensation (RMC), which consists of basic pay, basic allowance for housing (BAH), basic allowance for subsistence (BAS), and the tax advantage that accrues because those allowances are not taxed; and
- Special and incentive pays, consisting of enlistment bonuses, selective reenlistment bonuses (SRB), sea pay, and assignment incentive pay; retirement pay, including the retirement annuity for those who serve for 20 years or more, and the newer Thrift Savings Plan (TSP), health care, and the voluntary separation pay program.

[9]Under DOD's nomenclature, a high-quality enlisted recruit holds a high school diploma and scores above the median on the Armed Forces Qualification Test (AFQT), DOD's enlisted entrance test of cognitive aptitude.

The report recommended the most promising tools in the near term and also recommended actions for the long term.

Near-Term Recommendations

CNA first recommended that the Department of the Navy refrain in the near term from advocating increases to basic pay, military housing, BAH, health care, or retirement pay. However, it recommended that the Department of the Navy pursue initiatives to improve the effectiveness and flexibility of these compensation tools. Second, it recommended an expansion of special and incentive pays, specifically sea pay, SRBs, enlistment bonuses, and assignment incentive pay (AIP). It also recommended that the department begin use of TSP matching and the voluntary separation pay program.

Regular Military Compensation

The report argues that too little of today's cash pay can be targeted at specific populations, such as people in hard-to-fill occupations. CNA also found that cash pay is generally inflexible, in that it cannot be turned on and off to help force managers meet specific goals. A similar argument was made by the DACMC. Like the DACMC, CNA argues that the current pay table is not well suited to motivating performance. For these reasons the study recommends reducing the share of compensation devoted to basic pay.

The study also finds that the dependency differential for housing allowances is not consistent with the goals of the Navy's HCS. Like DACMC, it argues that the differential is not cost-effective, because married members are not more productive than singles. The study sees the annually adjusted housing allowance as a far less flexible and useful tool for dealing with cost-of-living variations than AIP, because AIP can be adjusted at any time and responds to relative preferences for locations as well as geographic variations in the cost of living.

Sea Pay

The study argues that sea pay promotes the strategic goal of assignment. However, both SRBs and AIP are likely more flexible and cost-effective, because they can be targeted. For example, SRBs are paid to those making retention decisions. In the case of sea pay, payments are made to every sailor who goes to sea, regardless of manning shortfalls. Because the payments are made regardless of an individual's preference for sea duty, the Navy may be paying significant economic rents to those who would go to sea even at much lower levels of sea pay.[10] The

[10]The term "economic rent" comes from economics and in the case of sea pay refers to the payment above what is required to induce a member to go to sea.

study argues that a sea pay that was market-based (like AIP) and that varied by occupation would be more cost-effective and flexible.

Selective Reenlistment Bonuses and Enlistment Bonuses

SRBs and enlistment bonuses support the Navy's retention and recruiting goals. The CNA study argues that they are consistent with the principles of flexibility and cost-effectiveness. The study also argues that both tools could be improved.

Today the Navy pays its SRBs through an initial partial lump sum followed by annual anniversary payments. The total size of the bonus is determined as a percentage of the sailor's basic pay at the time of reenlistment.

Because members highly discount future payments, it would be more cost-effective to pay the entire bonus as a lump sum and eliminate the anniversary payments. On the other hand, if anniversary payments are retained, they would be more useful as motivators of performance if the Navy based them on the sailor's grade at the time of payment, rather than at the time of reenlistment.

As for enlistment bonuses, the report argues that many enlistees are unaware of enlistment bonuses. This reduces the power of those bonuses as incentives and therefore waters down their cost-effectiveness. It recommends that the Department of the Navy provide potential recruits with more information about the magnitude of enlistment bonuses.

Assignment Incentive Pay

Although this program is still in a pilot stage, early indications suggest that AIP has the potential to be a powerful compensation tool and would support the Department of the Navy's strategic goals. The AIP pilot began in 2003. AIP levels are determined by a market-based system whereby sailors submit bids for the amount of pay they would require to volunteer for less-desired locations. Bids are constrained by limits set by the Navy for each location, and the limits can be changed as staffing shortages become more or less severe. The AIP system is an auction, with the winners being the sailors who submit the lowest bids. (The Navy can take other factors into consideration, for example, whether the individual is approaching a critical decision point.)

To date, empirical studies of the program are lacking. The report notes that key issues, such as gaming of the system, remain to be resolved. CNA finds that the program is consistent with the goals and guiding principles of the HCS.

Retirement Pay

Drawing from past studies, the CNA report concluded that the military retirement system is not flexible or cost-effective. For the near term the study recommended that the Navy seek out tools to offset the flaws of the system. Specifically, it recommended the Navy consider Thrift Savings Plan (TSP) matching and voluntary separation pay (VSP). For the longer term it recommended that Navy leaders support efforts toward military retirement reform.

Health Care Benefits

Like the DACMC, the report recommended increasing the share of health care costs borne by members. It also recommended that members and their families be required to decide whether they will use Tricare and which option they will use, to reduce uncertainty about who is using the military system and thus facilitate management of the system. The report found poor alignment between Tricare for Life and the strategic human resource goals of the Department of the Navy, because the costly benefit is deferred so far into the future that it has relatively little effect on recruiting and retention. CNA recommended working in the near term toward repeal of the benefit.

TSP Matching

The Thrift Savings Plan is a 401(k)-type plan established in the 1980s for federal civilian employees. In recent years military members have been permitted to participate, but the government does not fund matching contributions for them as it does for civilians.

A number of studies, including several study groups such as the DACMC and the earlier 2000 Defense Science Board Task Force on Human Resources, have recommended TSP with government matching. The CNA report recommended that in the near term the Department of the Navy begin to use TSP matching as a tool for achieving longer careers by matching the contributions of members with more than 20 years of service.

Voluntary Separation Pay

Voluntary separation pay is a force-shaping tool that can be targeted to mid-career personnel who would prefer to complete 20 years of service, to induce them to separate before reaching 20 years. The CNA report argues that VSP can help shape the force by facilitating the smooth transition of personnel from service, and that it is flexible because it can be targeted to individuals with specific occupations, years of service, or other desired characteristics. In the near term the report recommended using VSP to offset the lack of force management flexibility inherent in the military retirement system.

Longer-Term Recommendations

In the longer term the Navy is not constrained by the current operational environment. Furthermore, CNA argued that cooperation with other services and congressional approval are more feasible over the longer term than in the short term. Looking to the long term, the report's findings are consistent with those of the DACMC:

1. Compensation is too heavily weighted toward deferred compensation, such as retirement pay and health care benefits for retirees.
2. Compensation is too heavily weighted toward in-kind compensation, such as health care and housing.
3. Cash compensation is too heavily weighted toward basic pay and too little toward targeted pays, such as bonuses, sea pay, and AIP.

CNA's long-term recommendations are also consistent with the DACMC's. First, the retirement benefit should be revamped into a two-part system, one to provide for members' old age and the other to facilitate force management. The first part would be a defined-benefit or defined-contribution plan that vests at 10 years of service and begins payment at age 60. The second part could take the form of exit payments that begin at various career points and vary by occupation. Alternatively, it could take the form of career gate pay that would go to members who complete various career milestones. Such a retirement system could provide greater force management flexibility and reduce the proportion of compensation that is deferred.

Some of the CNA recommendations for housing allowances and health care could reduce the share of compensation devoted to in-kind benefits. In general, the report recommends that the military get out of the family housing business and continue its policy of privatizing the housing stock. Bachelor housing for junior members should be provided only when the benefits of acculturating and mentorship warrant the cost involved. The report also recommends introducing a cafeteria plan that would give members greater choice in their mix of cash versus in-kind benefits; for example, members without dependents could choose a less expensive health benefit and more cash in their pockets.

Over the longer term the CNA report also recommended shifting a greater share of cash compensation away from basic pay and into targeted special and incentive pays like SRBs and AIP. Like the DACMC, CNA recommends restructuring the pay table to a time-in-grade model. The report also recommends full support of AIP as a powerful compensation tool for the longer term.

CRITIQUE OF TWO PREVIOUS MANPOWER STUDIES

Criteria for Critique of Previous Manpower Studies

In assessing the recommendations of the DACMC report and the CNA report, this committee considered both their potential to meet the needs of a transformed force and the feasibility of their implementation. The committee believes that a transformed force will require manpower and personnel systems that facilitate force management, are flexible, are cost-effective, and are consistent with volunteerism.

Some recommendations will be harder to implement than others, because they require the concurrence of the other services, necessitate changes in law, or run counter to service cultures. Chapter 5 discusses the importance of a strategy for managing change.

Retirement System

The DACMC and CNA reports echo a long line of earlier studies that call for improving the military retirement system.[11] The need for reform is indeed compelling. The retirement system hampers flexibility in several respects. It is also inequitable to those who do not serve for 20 years, and it is inefficient.

The current system reduces flexibility in force management by creating an implicit contract with people. The implicit contract makes it hard to require them to leave between the 10th and 20th years of service, even if they are unproductive or if their services are no longer needed, as during a drawdown. Its cliff vesting and immediate annuity produce relatively uniform career lengths, even though more varied lengths would make more sense.[12] For example, the Navy might be better off if most fighter pilots left after 10 to 14 years and most medical personnel remained for 30 years or longer.

The 20-year retirement system also forces many officers to take up too many assignments in too few years. To acquire the necessary breadth of experiences needed over their careers, officers must serve in a variety of assignments, typically at increasing levels of command. Since Congress passed the Goldwater-Nichols Act of 1986,[13] those who hope to rise to the general officer or flag officer ranks

---

[11]A history of military retirement studies, commissions, and associated legislation, focusing on the past 60 years of proposed reforms, can be found in John Christian, 2006, *An Overview of Past Proposals for Military Retirement Reform*, RAND Corporation, TR-376-OSD, Santa Monica, Calif. An excellent assessment of the current retirement system, summarizing past studies, and a presentation of options for reform can be found in John T. Warner, 2006, *Thinking About Military Retirement*, Center for Naval Analyses, CRM D0013583.A1 Final, Alexandria, Va., January.

[12]"Cliff vesting" is a type of vesting that occurs entirely at a specified time rather than gradually. Until the specified time there is no vesting, at which point the benefit becomes fully vested (e.g., current military retirement plan after service of 20 years).

[13]Goldwater-Nichols Department of Defense Reorganization Act of 1986, 10 U.S. Code (2000) §§ 151 et seq.

must also serve for several years in joint assignments. But fitting so many assignments into a 20-year window means rotating through each posting quickly. As a result, many officers may not be getting the time they need to learn what they need to know from successive jobs.

The link between actions and consequences is sometimes broken, because individuals rotate out before having to experience the consequences of their actions. Breaking this link may encourage people to take actions that look good but are not actually productive. A longer officer career could permit longer assignments.

Today's system also limits the services' agility in dealing with rapid changes in manpower requirements. For example, in a downsizing the services can move quickly to lower the number of people brought in at entry level each year. But reducing accessions risks creating a hole in the force that cannot be filled later; a small entry cohort in one year translates into a small career cohort of future trainees, journey-level personnel, midgrade personnel, and ultimately senior personnel. A service can usually compensate for a single small entry cohort by recruiting larger groups in subsequent years. But a few consecutive years of reduced accessions creates "year group holes" and reduces the pool from which future leaders can be drawn.

An alternative way to deal with changing requirements is involuntarily separating midcareer personnel. To midcareer personnel with an eye on retirement, however, involuntary separation may appear to break the implicit contract under which they serve. During the 1990s, the DOD offered two programs, the voluntary separation incentive (VSI) and selective separation benefit (SSB), to smooth the transition for those who separated with less than 20 years of service.

Besides being inflexible, the retirement system is inequitable. Only about 10 percent of enlisted personnel and half of officers serve for long enough to become eligible for the retirement benefit. An alternative system would vest more members earlier, with benefits scaled to their lengths of service.

The system is inefficient because too much of military compensation is deferred. The DACMC and CNA reports both found that because members' discount rates[14] are high, the DOD could achieve the same retention patterns and therefore the same force at lower cost by shifting more of the total compensation dollar into immediate benefits. Estimates of the inefficiency in steady state suggest that the savings would amount to about $2 billion per year.[15]

---

[14]"Discount rate" is defined as the interest rate used in calculating the present value of expected yearly benefits and costs (see Appendix A, "Definition of Terms," in Defense Advisory Committee on Military Compensation, 2005, *The Military Compensation System: Completing the Transition to an All-Volunteer Force,* Arlington, Va., April).

[15]See Beth J. Asch and John T. Warner, 1994, *A Theory of Military Compensation and Personnel Policy,* RAND Corporation, Santa Monica, Calif.; Beth J. Asch and John T. Warner, 1994, *A Policy Analysis of Alternative Military Retirement Systems,* RAND Corporation, Santa Monica, Calif.; and John T. Warner, 2006, *Thinking About Retirement*, Center for Naval Analyses, CRM D0013583.A1 Final, Alexandria, Va., January.

The DACMC and the CNA studies outlined basic reform proposals, but neither offered specifics. Their general proposals are similar to each other and to the recommendations of other recent commissions and studies.[16]

Most of these studies call for a three-pronged approach to retirement reform:

1. Vesting more members to make the system more equitable;
2. Using transition or separation pay routinely, targeting particular occupations and beginning the payouts at varied career points depending on occupation, thus facilitating different career lengths for different occupations; and
3. Increasing base pay or adding a new career gate pay, to ensure retention through a less back-loaded and therefore more efficient system.

Estimates indicate that under such an approach the services could achieve the same retention patterns and experience mix at lower cost, more members would be vested, and the services would have greater force management flexibility.[17]

From the standpoint of feasibility of implementation, retirement reform has proven exceedingly difficult. Retirement benefits were trimmed in 1980 when Congress changed the basis for calculating retired pay from an individual's final year of basic pay to the average of his or her highest 3 years of basic pay. The Military Retirement Reform Act of 1986, also called Redux, made more substantial cuts. For members who joined after 1986, Redux would have reduced the initial annuity of members with 20 years of service (but not for those with 30 years). It would have lowered the cost-of-living adjustment between the point of military retirement and age 62 regardless of the number of years served. Redux was intended to shrink the number of service members staying in service for 20 years, while increasing the likelihood that a person who stayed for 20 years would continue to the 30-year mark.

Beginning in the late 1990s with recruiting and retention already challenged by a booming civilian economy, the post-1986 cohorts approaching retirement began to realize that their retirement system would be less generous than that of previous groups. Military and retiree associations lobbied for repeal, and military leaders told Congress that Redux hurt their ability to staff the force. In 2000 Redux was repealed for all practical purposes when Congress gave members the choice to go back to the pre-Redux plan.

The Redux experience is an object lesson in the obstacles to retirement reform and has the potential to put a chill on future reform efforts. Therefore, it

---

[16]John Christian. 2006. *An Overview of Past Proposals for Military Retirement Reform*, RAND Corporation, TR-376-OSD, Santa Monica, Calif.

[17]A final advantage is that such an approach could integrate the active and reserve retirement system under the same plan. See Beth Asch, James Hosek, and David Loughran, 2006, *Reserve Retirement Reform: A Viewpoint on Recent Congressional Proposals*, RAND Corporation, TR-199-OSD, Santa Monica, Calif.

is important to understand the lessons of Redux, why it did not work, and how to manage the change that a new system would entail. Chapter 5 discusses change management strategies for the Navy and Marine Corps.

Though favoring fundamental reform over the longer term, the CNA report recommended sidestepping such difficulties, and instead recommended greater use of TSP matching and VSP. The TSP match would vest after 20 years to motivate some members to stay in service longer. It is not known yet whether a TSP match would be cost-effective as a tool for inducing longer careers. The Department of the Navy needs to examine this issue further.

During the 1990s, the VSI and SSB programs were indeed effective in inducing members to leave prior to reaching 20 years.[18] In part this was because members knew at the time that if they didn't take the benefit and leave, they might be involuntarily separated. That said, VSP can be a valuable tool in downsizing, and the committee endorses its use.

VSP's potential effectiveness in shaping the force occupationally by inducing people in some career fields to depart between the 10-year and 20-year points will depend upon whether the Department of the Navy actually chooses to use the tool to encourage careers with more variable lengths of service. In the absence of fundamental retirement reform, the Department of the Navy may prove reluctant to break the golden handcuffs of the 20-year system and use VSP for this purpose. Thus, the committee believes that even though it will be difficult, the three-pronged approach of fundamental reform is crucial.

Special and Incentive Pays

The issues regarding special and incentive pays in general are multifaceted. The DACMC was concerned primarily with the sheer number of such pays, which that committee saw as a management issue. Another potential problem is that, with more than 60 distinct special and incentive pays, each controlled by its own legal authority, members may become confused about their pay. Of course, many members are never eligible for many pays (e.g., occupation-specific pays such as those for health professions).[19]

The DACMC was also concerned about the flexibility of special and incentive pays. The pays are often narrowly defined in legislation, each with a distinct set of rules for who is eligible for what amount of money and over what time period. Some of them (e.g., pilot bonuses, diving pay, and hostile-fire pay) have become

[18]Stephen Mehay and Paul Hogan. 1998. "The Effect of Separation Bonuses on Voluntary Quits: Evidence from the Military's Downsizing," *Southern Economic Journal*, Vol. 65, No. 1, July 1, pp. 127-139.

[19]Beth J. Asch, James R. Hosek, and Craig W. Martin. 2002. *A Look at Cash Compensation for Active Duty Personnel*, RAND Corporation, MR-1492-OSD, Santa Monica, Calif.

entitlements; their budgets are set annually, with little consideration of their contribution to force staffing or readiness. Other pays (e.g., career sea pay) give the Navy some discretion regarding eligibility, but only over a relatively narrow range of duties or occupational categories.

To address these issues the DACMC recommends consolidating all of the special and incentive pays into a few broad categories and permitting the services to allocate funds within a category based on staffing requirements rather than as entitlements.

The committee agrees with the need for consolidation and encourages the flexible use of special and incentive pays. A provision for annual adjustments for inflation to the total amounts provided for each broad category would help to preserve their usefulness and flexibility over time. History reveals a persistent temptation to develop new narrowly defined pays, however. To avoid the proliferation of new pays the committee recommends the establishment of rules to ensure that any new pays be fit into the broad categories.

The committee also worries that consolidation may encourage members in some occupations to argue that their pay should match that of those in other occupations whose special pays come from the same consolidated category. For example, doctors in different specialties might argue on equity grounds for the same special pay as other doctors. The service secretaries will play a crucial role in the implementation of a consolidation of special pay categories. The secretaries will have to perform careful analysis, exercise good judgment, and be firm managers.

Both the DACMC and the CNA study recommend increasing the relative share of cash compensation devoted to special and incentive pays. Both argue that pay in the private sector varies much more according to occupation than does military pay. The shift would give the services more flexibility to compete against the private sector by targeting resources to recruiting and staffing problems without increasing compensation budgets.

Those who oppose the targeting of pay according to occupation or performance argue that targeted pays fail to recognize the shared risk under which a unit operates and are detrimental to unit solidarity. That argument carries little weight, however. For decades higher-paid nuclear-qualified and lower-paid nonnuclear-qualified sailors have worked together aboard the same submarine with no visible impact on teamwork. Opponents also argue that paying more to people in some occupations is detrimental to good order and discipline, because a commander loses authority over his or her troops if the commander does not earn more than the troops do. Yet marine pilots salute their nonpilot superiors just as sharply as do their nonpilot peers, even though their aviation bonuses may mean that their paychecks are higher than the boss's. Past research shows that special and incentive pays are indeed effective in improving retention in critical skills and channeling recruits into hard-to-fill ratings.[20] The committee supports the shift toward a pay package in which a larger fraction is flexibly targeted.

The committee is concerned, however, that without changes in the retirement system, the services may fail to use larger special and incentive pools to manage career lengths effectively. For example, among the services, the Navy currently has the largest share of special and incentive pays.[21] While these pays are used to enhance paychecks in some occupations to solve crucial staffing problems, the services' retention patterns and career lengths tend to be similar by occupational area.[22,23] This suggests that the services do not use special and incentive pays flexibly to vary career lengths, but rather to produce similar career length. When retention falls, these pays are used to bring retention back into line so career lengths are restored. Without a changed retirement system there may be little incentive to use special and incentive pays more flexibly to manage career lengths differently.

Devoting a greater share of immediate compensation to special and incentive pays could reduce members' lifetime earnings, because these pays are not included in the calculation of retirement pay. Furthermore, to the extent that special and incentive pays are started and stopped, members may view their total compensation as being more uncertain. Uncertainty in their pay will make compensation less valuable, in an expected-value sense. To hold the expected value constant, basic pay might need to rise to offset the higher risk. Thus, the overall cost of shifting a higher proportion of cash compensation toward special and incentive pays is unclear.

The CNA report makes specific recommendations for changes to sea pay, SRBs, and AIP. CNA summarizes the literature on sea pay as follows: "The general findings are that sailors do react negatively to more sea duty; that is, retention falls if they expect more sea duty, longer tours, or are about to roll from shore to sea. And like all forms of pay, sea pay has a positive and significant effect offsetting this decline in retention."[24]

This committee strongly favors reforming sea pay to allow targeting by occupation. Such reform would give the Navy the flexibility to reverse manning shortfalls in key occupations without raising sea pay in occupations that are

---

[20]Michael Polich, James Dertouzos, and James Press. 1986. *The Enlistment Bonus Experiment*, RAND Corporation, R-3353-FMP, Santa Monica, Calif.

[21]Beth J. Asch, James R. Hosek, and Craig W. Martin. 2002. *A Look at Cash Compensation for Active Duty Personnel*, RAND Corporation, MR-1492-OSD, Santa Monica, Calif.

[22]U.S. Government Accountability Office. 2002. *Management and Oversight of Selective Reenlistment Bonus Program Needs Improvement*, GAO-03-149, GAO, Washington, D.C., November.

[23]Beth J. Asch and James R. Hosek. 2004. "Looking to the Future: What Does Transformation Mean for Military Manpower and Personnel Policy?" *The All Volunteer Force: Thirty Years of Service,* Barbara Bicksler, Curtis Gilroy, and John Warner (eds.), Brassey's, Washington, D.C., pp. 57-89.

[24]Michael Hansen and Martha Koopman. 2005. *Military Compensation Reform in the Department of the Navy,* Center for Naval Analyses, Alexandria, Va., December, p. 34.

already fully staffed. Implementing the new system as an AIP-like auction would save money while reducing the number of involuntary assignments to sea duty.

CNA's recommendations for the timing and calculation of SRB payments also make sense, but this committee is concerned that paying SRB as a lump sum rather than installments poses some risk of default on reenlistment contracts by sailors or marines who have already collected the full payment. The Navy and Marine Corps can limit defaults by pursuing such behavior aggressively. Calculating installment payments based on pay grade at the time of payment rather than at the time of reenlistment will increase the cost of SRBs somewhat, but the committee believes the improved incentive for performance will be well worth the cost.

Though all the evidence on AIP is not yet available, this program has the potential to be highly effective as a flexible and efficient compensation tool. The committee recommends early evaluation to identify areas where AIP might be improved. For example, the AIP cap may need to be increased for some especially hard-to-fill locales, such as Hawaii.

The committee also recommends extending the AIP concept to other areas of cash pay. The discussion above noted the potential value of changing sea pay to a system like AIP. Chapter 4 outlines a pilot program the Navy might use to evaluate an AIP-like sea pay. A similar idea could be implemented for the submarine and air communities. As discussed below, BAH might also usefully be replaced by AIP.

Health Benefits

Both the DACMC and CNA reports highlight an important inefficiency in the military health care benefit, namely, the crowding out of civilian benefits by Tricare for pre-age-65 retirees. While the premiums for health plans offered by civilian employers have grown dramatically in recent years, Tricare premiums have not been adjusted since 1995.[25] Because Tricare's cost-sharing arrangements are increasingly generous relative to those of most civilian employers, military retirees under 65 years old are turning in growing numbers to Tricare, even though their postmilitary employers offer a health benefit.

The net beneficiary of this shift is the civilian employer. According to the DACMC report,[26] a retiree with dependents who opts to use Tricare instead of employer coverage saves his or her civilian employer $7,000 a year. After accounting for copays, deductibles, and premiums, he or she saves $2,500 a year. In fact,

[25]Defense Advisory Committee on Military Compensation. 2006. *The Military Compensation System: Completing the Transition to an All-Volunteer Force,* Arlington, Va., April, p. xxvii.

[26]Defense Advisory Committee on Military Compensation. 2006. *The Military Compensation System: Completing the Transition to an All-Volunteer Force,* Arlington, Va., April, p. 78.

the DACMC reports that some employers actually pay cash bonuses to entice employees to shift to Tricare.[27]

To address this issue both the DACMC report and the CNA study recommend increasing Tricare fees for retirees under age 65 to be commensurate with the cost-sharing provisions in civilian employer plans. Research shows that individuals are responsive to the price they pay for health care services and prescription drugs. It also shows that copayments are cost-effective in encouraging wise health care spending. Increasing the share of costs borne by military retirees under age 65 would improve cost-effectiveness for DOD and could also reduce the allure of Tricare for members who have civilian health plans available to them.

To induce more cost-effective use of health benefits the committee favors some rise in the cost-sharing arrangements. Specifically, copays and premiums should be indexed to the annual cost-of-living adjustment to the military retirement annuity.

While shifting the cost of coverage from the government back to civilian employers makes good sense, the committee is concerned that asking military retirees to pay the full marginal costs of that shift may not be politically feasible. Retirees are well represented by powerful associations, which argue that those who serve for 20 years or more did so in the expectation that their lifetime health care would be free.

Solutions could be made more politically feasible if the costs of health care could be shifted back to civilian employers without raising substantially the costs to retirees. An alternative proposal is to require pre-65 retirees to stipulate a primary and secondary insurer and to give these retirees an incentive to name Tricare as the secondary insurer. That is, just as employers provide retirees with bonuses to choose Tricare, the Department of the Navy could provide retirees with a bonus to choose their civilian employer health benefit instead.

The DACMC also recommended that the pre-65 retirement annuity be funded on an accrual basis. This committee endorses that recommendation as a way to improve visibility into the future costs of policy changes. The committee agrees with the DACMC report that in the first year of the change, the top line of the DOD budget would need to increase to offset the effects of this accounting change.

For dependents of active-duty members Tricare has three options: an HMO choice known as Tricare Prime and two preferred provider organization (PPO) options—Tricare Extra and Standard. Tricare Prime requires active enrollment, but Tricare Extra and Standard require no enrollment commitment and can be used on a case-by-case basis for each medical need. For example, under Tricare Extra and Standard dependent beneficiaries can seek care at military treatment facilities on a space-available basis or receive care from civilian providers. CNA argues that the absence of enrollment makes it difficult for DOD to manage Tricare efficiently.

[27]Defense Advisory Committee on Military Compensation. 2006. *The Military Compensation System: Completing the Transition to an All-Volunteer Force,* Arlington, Va., April, p. xxvii.

The CNA report suggests requiring military families to decide whether they will use Tricare and which option they will use.

If enrollment is required, open enrollment windows will be needed to allow members to review their choices periodically. Members would also need flexibility to change plans even during closed periods as they rotate from one location to another or as other circumstances change. Given the potential for such turbulence, this committee is unable to assess the degree to which required enrollment would actually improve DOD's ability to manage the system. Moreover, it is possible that the level of choice inherent in today's system has a positive effect on morale and ultimately retention.

Pay Table Reform and Pay for Performance

Both the DACMC and CNA reports recommended restructuring the basic pay table from one based on grade and time in service to one based on grade and time in grade. This committee agrees that a pay table based on time in grade would give individuals who were promoted earlier a pay differential that was persistent and more substantial than one based on time in service. On average the present value of basic pay over a career would be about 3.9 percent higher for members receiving an early promotion under a time-in-grade table, compared with 1.9 percent higher under the current table. Because the Navy promotes personnel faster than average, the comparable figures for the Navy are 4 percent and 2 percent. Consequently a time-in-grade pay table would embed greater pay-for-performance incentives in the promotion system.

The committee also believes that a time-in-grade table could potentially accommodate lateral entrants more smoothly. Currently individuals entering with a civilian skill may enter at a higher grade, reflecting higher education, but at years of service 1 (YOS 1). A time-in-grade table that allows individuals to enter at advanced grades could provide entry-level compensation sufficiently high to make military service attractive. Furthermore, prior-service members who left service may consider returning to service if they could reenter at a higher grade. The current time-in-service table, where the clock stops, or even gets reset to the beginning (i.e., to YOS 1) can hinder the reentry of prior-service members with valuable skills.

These considerations suggest that a time-in-grade pay table is a good idea, because it would improve the monetary incentive for performance and could support more varied career lengths. The Navy should recognize that there are also other ways to enhance performance incentives. Chapter 2 describes career paths under which individuals might spend more time in a given pay grade. With such alternative paths promotion would not be the main reward for outstanding performance. An alternative performance evaluation system would need to be designed, linking pay to individual performance not just through the promotion system but also through periodic individual evaluation. The Department of the Navy should investigate an alternative personnel evaluation and incentive system to reward career paths that do not necessarily involve moving up the chain of command.

Housing

The committee agrees with the finding of the DACMC and CNA that today's differential in the BAH rate between those with and without dependents is inefficient, since members are not differentially productive based on their dependents' status. The committee also finds that the system encourages early marriages and higher rates of divorce relative to similar civilians according to the DACMC.[28]

The DACMC argues that eliminating this distinction makes sense, and the most practical approach to doing so will require paying the higher "with dependents" rate to those without dependents. The DACMC estimated that across DOD the annual cost of this change in 2005 dollars would be about $500 million. The added cost might be offset by improvements in recruiting and retention due to the increase in military compensation for the average member. Also offsetting the cost would be changes in the composition in the active force toward members without dependents; the more single force would have lower costs associated with health care benefit for dependents, DOD schools, permanent change of station, overseas cost-of-living-allowance, and other costs that increase with dependents. The DACMC estimated that the annual cost savings from these programs would be about $100 million in 2005 dollars. If recruiting improves, fewer recruiters would be needed, providing additional savings.

An additional recommendation that would lower incentives for junior enlisted members to form families would be to pay BAH, regardless of the member's housing arrangement, and charge those living in government housing fair market value for their housing. Junior enlisted living in barracks or onboard ships would receive a net increase in pay because the BAH rate they would receive would be greater than the fair market value for their quarters. This recommendation also makes sense because junior enlisted members would no longer have an incentive to get married in order to receive the BAH rate. For more senior personnel the move would have a neutral fiscal effect because those in government housing would be charged a rental rate equal to their BAH rate.

The CNA report recommends replacing BAH with AIP. This recommendation has much merit. AIP accounts for all cost-of-living differences associated with different locales, not just housing costs, and accounts for member preferences. Eliminating BAH is likely to be politically difficult because it involves

---

[28] Jeffrey S. Zax and David Flueck. 1995. "Marriage, Divorce, Income and Military Marriage Incentives," 1995 discussion papers in Economics, Center for Economic Analysis, Department of Economics, University of Colorado at Boulder, May. Available at <http://www.colorado.edu/Economics/CEA/papers95/.html>. Accessed on October 5, 2007.

coordination with the other services. The other services do not yet have AIP and a voluntary assignment system. Thus, they are not yet set up to eliminate BAH. The committee recommends that Department of the Navy leaders support the replacement of BAH with AIP.

The committee agrees with recommendations to consolidate family separation pay with other special and incentive pays related to deployment. Deployment pay should reflect the arduous nature of deployment and the potential for extra expenses, rather than arbitrary distinctions between members with and without dependents.

Quality of Life

Quality-of-life (QoL) programs encompass a diverse set of benefits, typically provided in kind as subsidized goods and services rather than in cash. They include recreation programs, day care centers, commissaries, family support centers, educational assistance, and programs to address issues related to spouse employment and children's education. These programs help maintain morale and mitigate the unique demands of military service, including the requirement that members and their families be on-call at all times, endure separations, and cope with frequent long-distance moves.

Given scarce resources, finding the right balance between cash and in-kind benefits is a challenge. On the one hand, a compensation system that will meet the demands for readiness and capability in an expeditionary military is not likely to be one that is founded on paternalism. On the other hand, in-kind benefits may improve staffing and readiness. For example, fitness centers might make sailors and marines more productive; tuition assistance might help the services selectively attract and retain enlistees who are smart and motivated to achieve; and high-quality child care centers might free members from on-the-job concerns about their young children.

In-kind benefits are also often less economically efficient than cash, in that they cost more to provide than members would be willing to pay for them. Yet in-kind forms of compensation may actually be more efficient than cash in some cases. For example, the Department of the Navy may be able to take advantage of economies of scale and purchase some benefits at a lower cost than if each sailor and marine bought them independently.

The DACMC recommended devoting more effort to measuring and evaluating the costs and effectiveness of QoL programs. It called for developing metrics with care, however, because the small scale and geographic heterogeneity of many of these programs can make it difficult to measure their effects on readiness.

One presentation provided to the committee from the Commander, Navy Installations Command, Fleet and Family Readiness, suggests that the Navy is now using metrics routinely to track QoL programs.[29] Such tracking is a very good start.

Unfortunately, however, it is often difficult based on such tracking to distinguish improvements that are sparked by deliberate policy changes from those that would have occurred anyway under the status quo policy. The committee recommends that the Department of the Navy also undertake more formal evaluations that involve baseline control groups and control periods. Such evaluations could help the services to assess the relative cost-effectiveness of various programs and improve understanding of the trade-offs among them.

The DACMC singled out spouse employment and children's education as areas of special interest. Increased choice through assignment, such as through the AIP, can help address these issues. In such areas as Hawaii where the quality of public schools makes assignments a particularly poor choice for families, AIP caps may need to be increased above the current limit, and other measures may need to be taken to increase the attractiveness of Pearl Harbor. The Navy has expanded construction of child care centers throughout the country. In areas where jobs are plentiful this step may help in the area of spouse employment, although it also further increases the benefits package for members with families relative to that of singles.

CNA also recommended introducing a cafeteria plan that would give members a choice between cash and a menu of qualified (nontaxable) benefits, including health care, child care, and life insurance. One way to implement this approach in the near term is to allow members to set up a flexible spending account where they could set aside pretax dollars for dependent care and other qualified expenses like health care. The committee believes additional research is needed to determine the benefits and costs of a military flexible spending account, taking into consideration the effects on the Department of the Navy's ability to attract and retain personnel to meet its requirements, and any offsetting costs associated with reduced economies of scale in the provision of in-kind benefits.

Reserve Compensation

The DACMC recommended that activated and mobilized reservists earn the same pay as active-duty members, that active and reserve compensation be coordinated in a systems approach, and that the reserve components be given a new flexibility to set bonuses by geographic area and unit to reflect the location-specific nature of manning reserve units. The first recommendation addresses equity concerns while the second two recommendations address force management and flexibility of compensation. All three recommendations support the compensation

[29] John Baker, Head, Fleet and Family Readiness Directorate, Navy Installations Command, "Navy Family Center–Manpower and Personnel Benefits," presentation to the committee, November 14, 2006, Washington, D.C.

goals and so seem sensible in terms of the supporting total force objectives of the Department of the Navy.

Followed blindly, the first recommendation can lead to policies that run counter to the Department of the Navy's compensation goal of cost-effectiveness. For example, active and reserve members have different retirement systems. Several proposals before Congress have recommended changing the reserve retirement system to mirror the active system, including lowering the age at which reservists can begin earning retirement benefits, which is currently 60. Research shows that such a policy would increase cost but would do little to improve retention.[30]

The committee generally favors changes that would bring the immediate cash pay of reservists serving on active duty in line with that of their active-duty counterparts. In some instances, however, the committee finds that compensation differences make sense in light of the unique missions of the two components. The committee strongly supports location-specific bonuses for reservists.

The committee finds that changes to the reserve retirement system should generally be made in the context of overall military retirement reform. Proposals for changing reserve retirement outside of such a context should be examined on a case-by-case basis to determine the likely costs as well as the potential contributions toward Department of the Navy human resource goals for the active and reserve components.

Studies of the Effects of Deployment on Service Members

Long and frequent deployments associated with the wars in Iraq and Afghanistan are raising concerns about the relationships among work pace, stress on the force, morale, and retention. Recent research provides some insight, but the Department of the Navy needs additional information.

Past research indicates that individuals' attitudes toward service worsen when the amount of expected deployment differs from preferred amounts. Evidence from the 1990s shows that deployment has a positive effect on reenlistment for most military members: reenlistment is higher for personnel with some deployment than for personnel with no deployment. However, when individuals reach three or more deployments, reenlistment probabilities drop somewhat.[31] A similar result is found for officers. Past studies of the Navy find that the more sea duty a sailor expects in his or her next term, the less likely he or she is to reenlist or

[30]Beth J. Asch, James R. Hosek, and David S. Loughran. 2006. *Reserve Retirement Reform: A Viewpoint on Recent Congressional Proposals*, RAND Corporation, TR-199-OSD, Santa Monica, Calif.

[31]See James Hosek and Mark Totten, 1998, *Does PERSTEMPO Hurt Reenlistment? The Effect of Long or Hostile PERSTEMPO on Reenlistment*, RAND Corporation, MR-990-OSD, Santa Monica, Calif.; James Hosek and Mark Totten, 2002, *Serving Away From Home: How Deployments Influence Reenlistment*, RAND Corporation, MR-1594-OSD, Santa Monica, Calif. A similar result is found for officers; see Ronald Fricker, 2002, *The Effects of PERSTEMPO on Officer Retention in the U.S. Military*, RAND Corporation, MR-1556-OSD, Santa Monica, Calif.

extend. However, increases in sea pay can offset negative retention effects of sea duty.[32] Heidi Golding finds that a $50 increase in total monthly sea pay results in an 11 percent increase in the completion of 48-month sea tours.[33]

Only one study has looked at information post-9/11, and this study is based on focus group and survey data from all services and from both officer and enlisted personnel.[34] The study provides information on retention intentions only and not on actual retention behavior. Still, it is suggestive of some initial insights.

The study finds that active-duty members who report higher than usual work stress have a higher intention to stay in the military, suggesting that these individuals are engaged in meaningful work that, though demanding, is satisfying. Specifically, those who had the usual amount of work stress had a 22 percent likelihood of stating they were likely or very likely to stay in service versus 43 percent of individuals who stated they had much more than usual stress. On the other hand, the survey data show that more individuals are working longer than the usual duty days, among both deployed and nondeployed personnel. In addition, they are under more than the usual work stress, and the survey data indicate that those who work longer days have a lower intention to stay. It appears that the negative effect of longer days is not sufficiently negative because, as mentioned, those under more than usual work stress have a higher intention to stay. This suggests that those who are working more frequently under unusual stress find that the work is sufficiently gratifying, perhaps because it supports a meaningful mission, that it offsets the negative effects of longer days on the intention to stay. The study also finds that unusual work stress is lower when members feel that they and their units are well prepared.

The results of this study indicate that increases in unusual work stress and long work hours interact in complex ways with intentions to stay. The Department of the Navy needs to understand how unusual stress levels, work pace, and deployment schedules are affecting morale, performance, and retention.

FINDINGS AND RECOMMENDATIONS

The following findings and recommendations are those of the committee based on its review of the prior studies.

[32]Heidi Golding with David Gregory. 2002. *Sailors' Willingness to Complete Sea Tours: Does Money Matter?*, Center for Naval Analyses, CNADRM 00006886/Final, Alexandria, Va., December.

[33]Heidi Golding with David Gregory. 2002. *Sailors' Willingness to Complete Sea Tours: Does Money Matter?*, Center for Naval Analyses, CNADRM 00006886/Final, Alexandria, Va., December.

[34]James Hosek, Jennifer Kavanagh, and Laura Miller. 2006. *How Deployments Affect Service Members*, RAND Corporation, MG-342-RC, Santa Monica, Calif.

Deployment and Pace of Work

Finding: Recent and planned changes in Navy and Marine Corps missions, unit structure, and materiel will likely have adverse effects on people and readiness.

Studies based on surveys and focus groups indicate that whether service members are deployed or at home, their morale and intention to continue in service appear to suffer when work hours are unexpectedly onerous, and when the realities of military life differ substantially from the individuals' expectations. The wars in Iraq and Afghanistan, other worldwide deployments and commitments including the global war on terror, and the continual downsizing of the Navy pose the potential for adverse effects on people and readiness. While such effects may ultimately be ameliorated by the introduction of labor-saving technologies, they are exacerbated when workloads increase because the fielding of or training on new equipment lags behind the personnel reductions. Moreover, the challenges of irregular warfare, the continued shifting of shore work to contractors, and the Navy's plans to move from a pyramidal to an oval workforce structure (discussed in Chapter 2) are likely to lengthen work days for many sailors and marines. If individuals are overloaded beyond their ability to support the mission, both mission readiness and future retention are likely to suffer.

Recommendation: The Navy and Marine Corps need better evaluation programs to measure and interpret the effects of changes in workload, separation from home, length and repetition of deployments, and similar factors on readiness, morale, and intention to serve. The Assistant Secretary of the Navy for Manpower and Reserve Affairs, the Chief of Naval Personnel, and the Deputy Commandant of the Marine Corps for Manpower and Reserve Affairs should develop new metrics for the early identification of morale and retention signals related to deployment cycles, workload, and manning levels. Additional studies should be undertaken to understand better how deployment cycles, workload, morale, intention to continue, and actual continuation are related.

Compensation and Retirement Reform

Finding: The current package of immediate compensation is overly complex, lacks flexibility, is not conducive to Navy and Marine Corps force management, and generally costs more than its value to service members.

Recommendation: The Secretary of the Navy, the Chief of Naval Operations (CNO), and the Commandant of the Marine Corps (CMC) should support the following DOD-wide policy changes:

- Improve incentives for performance by reconfiguring the basic pay table to set pay based on grade and time in grade rather than time in service.
- Equalize the basic allowance for housing (BAH) rate for all service members, regardless of family status.
- Pay BAH to all service members and charge rent to those in government housing, with the rent equal to the fair market value of their housing.
- Consolidate all deployment-related pays, including the family separation allowance. Deployment-related pay should be set to reflect the nature of deployment.
- Consolidate the special and incentive pays into a much smaller number of categories, and work actively to help shape the categories to make these pays as flexible and useful as possible. In doing this, give particular attention to assignment and career incentive pay and selective reenlistment bonuses to best match the interests of sailors and marines with those of the services. Assuming that the broader pays are instituted, naval leaders, including personnel and budget offices, should make every effort to avoid "cost creep" as individual constituencies lobby for pay increases.

The CNO and CMC should direct that rigorous evaluations of the services' quality-of-life programs be carried out to understand their effects on readiness and ensure their cost-effectiveness. The leaders should also conduct research to assess the costs and benefits of flexible spending accounts for these programs, including health care.

Finding: The current military retirement system impedes flexible force management and is inequitable and inefficient. In addition, health care benefits for military retirees are not cost-effective.

Recommendation: The Secretary of the Navy (SECNAV), the Chief of Naval Operations (CNO), and the Commandant of the Marine Corps (CMC) should advocate and support changes to the compensation system that would accomplish the following:

- Encourage longer careers for some personnel and shorter careers for others.
- Offer greater flexibility in management of active and reserve forces.
- Result in more personnel who are vested.
- Generally provide more compensation up-front and less deferred to retirement.
- Improve integration of active and reserve forces.

A three-part system seems the most promising in terms of effectiveness and political feasibility. Such a system would consist of the following:

- An entitlement that allows members to accumulate benefits for old age,
- A career management tool for the services, and
- Additional front-loaded compensation that would also serve as a management tool.

The SECNAV, CNO, and CMC should also support a proposal that would require retirees under age 65 to stipulate a primary and secondary insurer, and that would give these retirees an incentive to stipulate Tricare as the secondary insurer. To induce more cost-effective use of health benefits, Tricare fees should be indexed to the annual cost-of-living adjustment to the military retirement annuity. Finally, the financing of the under-65 health benefit should be through an accrual fund, as it is for retirees over 65.

Finding: The reserve compensation system does not support the new role of the reserves as an operational force, nor does it consistently support the unique features of the reserves.

Recommendation: In general, the Department of the Navy should support policies that provide activated reservists with the same compensation as active-duty members, in equally ranked groups. But this should not interfere with other goals of providing compensation such as recruiting, staffing, achieving cost-effectiveness, and maintaining flexibility. The Department of the Navy should support the creation of a retention bonus that can depend on location or unit, reflecting the fact that reserve units draw manpower from local labor markets.

4

The Role of Research Tools in Implementing Change

In previous chapters the committee has said that large forces at work in and on the Navy and Marine Corps will alter the demands on sailors and marines and require changes in the manpower and personnel systems of the two services. Specialization and generalization seem to be pulling at the same time on future technicians and leaders. Navy and Marine Corps requirements for a transformed force demand new manpower and personnel policies that are more flexible and adaptive. Research tools ranging from surveys and analyses of administrative data to pilot demonstrations and experimentation have always been key tools in formulating policies that support military manpower transformation. They should be used now, too, as the naval forces strive to adapt manpower policies and procedures to the new demands of the transformed naval services.

The Navy must begin to prepare for manning newly constructed ships with smaller crews of sailors with new skills in new combinations, while Navy managers have to continue to man the existing fleet largely as it has in the past, with sizable crews of sailors who have a conventional range of occupations. At the same time, the introduction of new technologies and the downsizing of the Navy demand that more capabilities be available from a smaller yet increasingly talented, educated, and integrated workforce. As a result of the ongoing war in Iraq and Afghanistan, both the Navy and the Marine Corps have growing requirements for more specialized, highly skilled teams and individuals supporting the high demand-low density commands (e.g., Seabees, combat engineers, Special Forces, force reconnaissance, military police, computer and network specialists, and foreign area officers). Marines today engage in stability and security operations, and then transition to counterinsurgency operations. Skill requirements not previously dominant in the force range from language and cultural competences for both marines and sailors to the train-to-qualify concept, in which sailors must step aboard ship essentially ready to stand watch and carry out the duties of the positions to which they are assigned.

Achieving transformation will be complex, and the road to transformation may present some countercultural challenges. Establishing and sustaining transformation in Navy and Marine Corps manpower and personnel policies will require the use of a variety of research tools to address the many facets of the naval services' transformation equation. The external and internal forces that affect the ability of the Navy and Marine Corps to shape their manpower processes, investments, and personnel inventory will require that experiments and pilot demonstrations be conducted when appropriate, that surveys and analyses of existing data sources continue to be conducted on a routine basis, and that funding, forces, and equipment be made available to support an ongoing research agenda.

When data are not available to provide a body of knowledge from which to analyze diverse workforce configurations, organizational structures, or new skill sets, experimentation has been one of the research tools that the Department of Defense and the Navy in particular have used in achieving changes to the shape of military forces. Through experimentation, military planners learn the characteristics of proposed battlefield systems and become able to predict the performance of those systems in actual situations.[1] Lessons derived from evaluation of these research projects must be harvested if the Navy and Marine Corps are to meet the challenges of the future discussed in this report.

This chapter lays the groundwork for incorporating various research methods into transformation planning. It discusses the fundamentals of surveys, simulations, pilot demonstrations, the analysis of administrative data, and experimentation. Some of the findings and recommendations of previous military manpower research programs that had an impact on subsequent policy and implementation decisions are then presented. This chapter suggests specific strategies that the Navy and Marine Corps could use to address recommendations in this committee's report.

ROLE OF SURVEYS, SIMULATIONS, ANALYSIS OF ADMINISTRATIVE RECORDS, PILOT PROGRAMS, AND EXPERIMENTS IN IMPLEMENTING CHANGE

Military planners have a long history of designing various types of experiments, pilot programs, simulations, and surveys in order to learn the characteristics of proposed battlefield systems and become able to predict the performance of those systems in actual operational and combat conditions. With respect to manpower and personnel policies, there is also a large body of research conducted

[1]National Research Council. 2004. *The Role of Experimentation in Building Future Naval Forces*, The National Academies Press, Washington, D.C.

for the Navy using surveys and administrative data. In the context of this study, however, surveys and retrospective data may not always be reliable predictors for the transformation of manpower and personnel policies, and experimentation may well become a critical enabler of fundamental change when evaluating manpower and personnel policies.[2]

The discussion below presents five approaches to evaluating proposals for policy changes, reviews some of the relevant literature that explores the structure of well-designed studies (experiments, simulations, surveys, and pilot programs), and discusses the resulting advantages and disadvantages of the approaches. The pros and cons of each approach must be weighed by the analyst, and sometimes multiple approaches can be used.

Surveys

A survey is a system for collecting information to describe, forecast, compare, or explain knowledge, attitudes, and behaviors. Surveys involve setting objectives for information collection, designing research, preparing a reliable and valid data collection instrument, administering and scoring the instrument, weighting and analyzing data, and reporting the results. The Office of the Undersecretary of Defense (Personnel and Readiness), as well as the individual services, regularly survey, both civilian and military personnel in a number of areas. A good survey has the following features:

- Reliable and valid survey instruments,
- Sound research design that enhances the precision of the results,
- Sound choice of population or sample to reduce bias,
- Specific objectives that can be measured,
- Appropriate analysis plan leading to an accurate reporting of survey results and logical conclusions, and
- Reasonable resources to conduct a statistically valid survey.[3]

For instance, two manpower factors sometimes considered for a survey are those that reveal attitudes and preferences about existing personnel and compensation issues, as well as proposed changes to these policies. Surveys are not reliable for getting good earnings data because people have poor recall. It is even

[2]The Collaborative International Dictionary of English v.0.48 (http://www.dict.org/bin/Dict) defines "experiment" as "a trial or special observation made to confirm or disprove something uncertain; esp., one under controlled conditions determined by the experimenter; an act or operation undertaken in order to discover some unknown principle or effect, or to test, establish, or illustrate some hypothesis, theory, or known truth; practical test."

[3]Arlene Fink. 2002. *The Survey Handbook, 2nd Edition,* Sage Publications, Thousand Oaks, Calif.

questionable how useful survey data are for predicting the effects of manpower and personnel decisions on transformation.

Simulations

A simulation develops a model of a real system and conducts experiments using the model for the purpose of understanding the behavior of the system or evaluating various strategies for the operation of the system. Simulations are invaluable for understanding both the effect of a policy change and the path of change over time. Simulations are particularly valuable when the outcomes of a proposed policy change will take time to unfold. In a prospective simulation, outcomes in the presence of the intervention (e.g., a policy or new program that is not consistent with routine procedures) are not measured directly but are projected on the basis of prior information or assumptions about the effects of the intervention or its elements. For example, in 1994 Warner and Asch used a computer simulation to look at military retention, cost, and productivity for each of a variety of retirement and personnel systems.[4]

Analysis of Administrative Records

Administrative records offer a solution for generating timely statistics at much lower costs than survey data collection. For the purposes of evaluating the effectiveness of military pay and personnel policies, administrative data provide information on the effects of military compensation on enlistments, attrition, and retention. Advances in computing technology and record-linkage accuracy have significantly increased the feasibility of expanded uses of administrative records for statistical purposes that avoid repetitive and burdensome inquiries of the population under study. These advances, coupled with spiraling costs of and resistance to traditional data collection, have increased the opportunity for significant benefits through use of administrative records in data collection, estimation, and evaluation systems.[5]

Pilot Programs

Pilot programs (often referred to as demonstration programs or trials) are a good way to test and fine-tune a proposed policy change when a controlled experiment is not feasible. They are valuable for developing and improving tech-

---

[4]Beth Asch and John Warner. 1994. *A Policy Analysis of Alternate Military Retirement Systems*, RAND Corporation, MR-465-OSD, Santa Monica, Calif.

[5]Ron Prevost and Charlene Leggieri. 1999. "Expansion of Administrative Records Uses at the Census Bureau: A Long-Range Research Plan," presentation at the Meeting of the Federal Committee on Statistical Methodology, Washington, D.C., November.

niques and demonstrating the effectiveness of specialized methods. They typically represent policy initiatives that are to be tested and examined on a limited basis before full-scale implementation.

A pilot intervention may be an entirely new program, but it is more commonly a significant modification to an existing program. Three distinguishing features of a demonstration lead to evaluation strategies that differ from those for ongoing programs:

- The intervention is new. In principle, evaluation activities can begin at the same time as implementation of the demonstration, or even before.
- The intervention has not been mandated by law for the entire program or service population.
- The intervention is applied to a restricted number of participants. During the relevant periods, some potential targets will be subject to the intervention and some will not.

However, pilot programs have limited scope, calling into question the advisability of generalizing their results. Pilot programs do not randomize treatment and control groups as a well-designed experiment would, so they do not produce findings as statistically powerful as do experiments.

Experimentation

Evaluating programs with experimental design methods has been one of the keys to defense planning in an era of high uncertainty and rapid technological change. An experimental design is appropriate and ideal for long-term changes that are transformational, where a long time horizon is feasible, and where the investment in doing the experiment is worthwhile. As a key tool in the shaping of military forces, this type of experimentation has traditionally provided a means for answering questions in order to operate effectively against future threats.

In a controlled experiment, program targets are randomly assigned either to an experimental (or treatment) group that will be subject to the program being assessed, or to a control group from which the program will be withheld. The program's impact is then estimated by comparing the average outcomes in the experimental group, after sufficient exposure to the program, with control group outcomes measured at the same time.

Because the experimental and control groups differ at the outset only by chance, they are considered fully alike at that point—equivalent, in the statistical aggregate, on all permanent and transitory characteristics. Subsequently the only systematic difference between the groups is exposure to the program. It is credible to infer that any postprogram differences between the two groups are caused by the program, provided that the differences are greater than what might occur by chance.[6]

Controlled experiments put policies into practice on a trial basis to provide direct evidence of their worth. The results are more credible because the policies are tested in practice, not by opinion polls or other indirect methods. These kinds of experiments can be carried out in such a way as to control the other factors that might otherwise tend to confound the cause-and-effect relationship between treatments and outputs.

The committee has chosen here to dwell a bit on the construction of experimentation. This is because Navy manpower and personnel transformation is calling for new skills and approaches to staffing that must coexist with traditional staffing for many years to come. As a result, indirect approaches to analyzing data (i.e., using administrative records or conducting surveys and simulations) may not provide policy makers with adequate information to evaluate the effect of these sometimes profound changes.

Experiments are developed by assembling a multidisciplinary team of advisers who possess the appropriate domain knowledge (practitioners as well as theoreticians), technical knowledge (for instrumentation and exercise support), and experimentation expertise (in design, measurement, and analysis), along with constituents who will have a stake in the outcomes. A conceptual model is then developed that embodies the key elements of the theory, formulation of questions and hypotheses, and collection of evidence and analysis. An experiment can be replicated in another setting when attention is paid to the principles of science that prescribe how these activities should be conducted, how peer reviews should be executed, and when attention should be paid to the widespread dissemination of findings and conclusions.[7]

Guidelines for good experimental designs should include the following:

- Objectives of the experiment that are clearly stated;
- Treatments (interventions) that are prescribed in detail;
- Treatment and control groups that are representative of the target population and are set up to assure valid estimates of the treatment effects;
- Size and duration of the experiment that are sufficient to meet the experimental objectives;
- Provisions for gathering reliable, comprehensive data and developing the analysis plan that are established in advance of the experiment; and

---

[6]William L. Hamilton and Peter H. Rossi. 2002. *Effects of Food Assistance and Nutrition Programs on Nutrition and Health: Volume 1, Research Design*, Food Assistance and Nutrition Research Report Number 19-1, U.S. Department of Agriculture, Washington, D.C., February, p. 3.

[7]Paraphrased from Richard E. Hayes, 2006, "Analytical Rigor in Joint Warfighting Framework," presentation at Bringing Analytical Rigor to Joint Warfighting Experimentation Conference, Joint Forces Staff College, Norfolk, Va., October 3-5.

- Performance metrics that are agreed upon by the experimenters and the policy makers who will be using the information.[8]

Articulating the benefits anticipated from an innovation or intervention and the conditions under which they are anticipated is important. You do not know for sure whether a process or policy will yield the anticipated results until you conduct the experiments. It is much better to realize problems and unintended consequences of the design during an experiment than it is to go out and just implement a new policy based on anecdotal evidence or some academic sense of what is right or wrong.[9]

Military leaders are particularly sensitive to the fact that the treatment group in an experiment may be provided with special privileges, raising questions of inequity. Given the magnitude of the costs associated with adopting ineffective personnel policies in the military and the foregone opportunities to invest these resources in other ways, the military has an obligation to evaluate new programs as carefully as possible before they are implemented. Thus, the decision not to experiment also involves ethical considerations.

There is an investment associated with experimentation, but the investment is modest with respect to the payoff in terms of improved military effectiveness and efficiency.[10] By avoiding funding sinkholes the cost more than justifies the expense.[11] Consider, too, that many of the costs associated with experimentation are start-up costs that new programs would have to bear anyway. Efficiencies realized in putting the policy into practice on a trial basis may lead to reduced costs in implementing a program later.

The end of this section describes an experiment in 1973 to test the feasibility of shortening the term of enlistment in the reserves. The end results demonstrate that there are best practices as well as risks for developing and executing an experiment. The risks associated with experimentation include:

- Moving ahead without sufficient evidence and understanding,
- Prematurely settling on an approach,
- Failing to capitalize on the creativity present in the force, and
- Progressing by trial and error rather than by theory.[12]

---

[8]Paraphrased from Gus W. Haggstrom, 1975, "The Pitfalls of Manpower Experimentation," The Rand Paper Series, RAND Corporation, Santa Monica, Calif., April, p. 4.

[9]ADM Harold W. Gehman Jr., USN. 2004. "The Role of Joint Experimentation," *Defense Horizons,* Number 46, Center for Technology and National Security Policy, National Defense University, December.

[10]An estimate for all experimentation and battlefield exercises is approximately 1 percent of the annual defense budget.

[11]Andrew Krepinevich. 2002. *Lighting the Path Ahead: Field Exercises and Transformation,* The Center for Strategic and Budgetary Assessments, Washington, D.C., February 25.

[12]David S. Alberts and Richard E. Hayes. 2005. *Campaigns of Experimentation: Pathways to In-*

Relating experimentation to transformation, Alberts and Hayes discuss the process of systematic observation, experimentation, and analysis that form the core answers to specific questions about why and how. In *Campaigns of Experimentation: Pathways to Innovation and Transformation*,[13] they argue that experiments are a source of knowledge, and that experimentation is a process rather than just a collection of experiments—a process that combines and structures experimental results and steers future experimentation activities.

Beyond the methodologies discussed above, there are other ways to gather information and evaluate the impact of proposed policy changes. The opinions of experts and the views of focus groups also provide indirect evidence useful in evaluating the potential effects of personnel policies. However, although the data are often compelling, the results can be difficult to replicate on a large scale. Even so, the information gathered from these approaches is valuable for ensuring the development of the right questions for a survey instrument, or establishing metrics for understanding a new policy rollout; they are good tools as well for helping decision makers fine-tune proposals and build momentum for change. But they provide no reliable base line of comparability. Consequently, the lessons derived cannot be scaled for a force-wide rollout, and give little indication of sustainability.

PAST FINDINGS, CONCLUSIONS, AND RECOMMENDATIONS DEVELOPED IN PREVIOUS MILITARY MANPOWER EXPERIMENTATION

Experiments on Enlistment

The Navy Enlistment Marketing Experiment was initiated in 1978 to examine the marketing effectiveness of the U.S. Navy recruiting program and to quantify the relationship between marketing efforts and enlistment achievements. The conclusions led to the reallocation of the advertising budget, development of a tenure-based recruiter incentive system, and a change in emphasis from recruit accessions to contracts. Some of the findings of this experiment included the following:

- The number of recruiters had a significant impact;
- Advertising expenditures of certain types were effective, while others were not;

novation and Transformation, DOD Command and Control Program, Pentagon, Washington, D.C., March.

[13]David S. Alberts and Richard E. Hayes. 2005. *Campaigns of Experimentation: Pathways to Innovation and Transformation,* DOD Command and Control Program, Pentagon, Washington, D.C., March.

- Socioeconomic factors such as unemployment and urbanization had major impacts; and
- Navy marketing efforts led to an expansion of the total market for military enlistments in addition to their impact on Navy enlistments.[14]

The Enlistment Bonus Experiment was initiated in 1981 in response to the challenges for defense managers to attract a sufficient number of new enlistees into critical occupational specialties within a reasonable level of recruiting expenditures. The design of the experiment quantified the effects of bonuses on increasing the total number of people joining the Army, shifting enlistments toward those with skills eligible for the test bonuses, and changing enlistees' choices among 4-year, 3-year, and 2-year terms of obligated service. The results showed that:

- An $8,000 bonus for 4 years had the potential to produce 4.1 percent more high-quality Army enlistments than did the control programs;
- A choice of an $8,000 bonus for 4 years or a S4,000 bonus for 3 years produced 5 percent more high-quality enlistments; and
- The experimental results showed that bonuses are a very flexible policy tool. Without altering the fundamental structure or level of military compensation, bonuses can be quickly altered when shortfalls appear in specific personnel categories.[15]

The Army's experiment in 1973 to test the feasibility of shortening the term of enlistment in the reserves is a demonstration of the hazards of conducting a manpower experimentation study without taking precautions to assure the validity of the experimental results. A small-scale controlled experiment began with Air Reserve Forces during which a few selected units were permitted to offer potential recruits a "3 × 3" enlistment option—3 years of regular reserve duty followed by 3 years in the Individual Ready Reserve (IRR). Other units were permitted to offer a "4 × 2" option—4 years of regular reserve duty followed by 2 years in the IRR. Seven months later the experimental results indicated that the shortened enlistment options proved to be far less effective in attracting new recruits than the military had expected. Unfortunately a month after the Air Force experiment began the Army Reserve Components undertook a similar experiment except that they offered the 3 × 3 option in all reserve units in 16 states and the 4 × 2 option in 12 other states. The Army officials were convinced that a shorter enlistment tour was essential for manning the reserves, and were concerned that a delay for

[14]Vincent P. Carroll, Ambar G. Rao, Hau L. Lee, Arthur Shapiro, and Barry L. Bayus. 1985. "The Navy Enlistment Marketing Plan," Warrington College of Business, University of Florida, Vol. 4, No. 4, Gainesville, Fla.

[15]J. Michael Polich, James N. Dertouzos, and S. James Press. 1986. *The Enlistment Bonus Experiment,* RAND Corporation, R-3353-FMP, Santa Monica, Calif., April.

experimentation would contribute to a further decline in reserve strength. Because these options did not stimulate recruiting enough to offset the man-year losses, the net effect of offering the options saddled the Army reserves with a large group of short-term enlistees. The Army also suffered a shortage in reserve strength for a substantial period as a result.

Given the magnitude of the costs associated with adopting ineffective personnel policies in the military and the foregone opportunities to invest these resources in other ways, the naval services have an obligation to evaluate new programs as carefully as possible before they are implemented. The above examples lay the groundwork for several proposed experiments that are based on the findings of this committee.

The Navy is familiar with the challenges and rewards of experimentation. The next section discusses a well-executed Navy experiment, Sea Swap, which was initiated in 2002 in anticipation of changing policy to get more productivity out of each hull by forward-basing ships and swapping multiple crews into them. The Navy developed inputs, outputs, and metrics in advance of the experiment and evaluated and assessed outcomes with a control group of ships on regular deployments.

The Navy Understands the Value and Challenges of Experimentation

As stated earlier, the naval services have been willing to undertake important projects and trials to determine whether a new method of operating would improve productivity, combat effectiveness, and quality of life. In some cases these projects have been intended to explore whether changes to existing operating procedures would improve operations. The most recent major trial of that kind is the Navy's Sea Swap project.

The Navy began manning ballistic missile submarines with two crews each in the 1960s. These crews, identified as Blue and Gold, rotated aboard the ships, each taking charge from the other for a strategic missile patrol; the off-duty crew devoting itself to training and administration in homeport during the patrol period served by the other. This continues in the Ohio-class ballistic missile submarines in service today.

In the 1990s the Navy's Mine Warfare Command confronted the need to provide mine countermeasures ships to the 5th Fleet Commander in the Arabian Gulf. The great distance (and relatively slow speed of the ships) dictated that the ships, once deployed, should stay on station if possible. Political and economic conditions in the region, however, made it impractical to home-base crews and families in Bahrain, where the ships would be. The Navy first tried and then adopted the procedure of forming four crews for groups of three ships, rotating the crews through rest and training, maintenance, ship training, and forward deployed ships as they moved from hull to hull, from Ingleside, Texas, to Manama, Bahrain, and back.

In two other cases the Navy created two crews per ship, Blue and Gold style. The patrol coastal (PC) ships, also stationed in Bahrain, support Special Forces operations with small Navy special warfare combatant crews, two rotational crews per ship. Two crews have served also in one of the Navy's experimental ships, high-speed vessel (HSV)-2 Swift, to permit that ship to remain underway during intensive periods of hull-form testing at sea.

Sea Swap was undertaken during 2002-2005 to see whether the approach to ship manning that had worked well in the rarified resources realm of the strategic missile fleet, and with the small crews of (relatively uncomplicated) mine warfare ships and PCs, could be employed to extend the on-station time of Spruance-class destroyers and Arleigh Burke-class guided missile destroyers.

This project was judged by the Navy and the Center for Naval Analyses to be a qualified success. The Government Accountability Office (GAO), while indicating that the project had taught the Navy many valuable lessons, criticized certain aspects of the test.[16] The GAO cited issues of data collection and methods of measurement in recommending improvements in the trial's technique and questions for further examination.

Sea Swap consisted of two kinds of crew exchanges. In the first the Navy took advantage of the fact that several modernized Spruance-class destroyers in the Pacific Fleet were being decommissioned. These ships would, when put out of service, make crews available for additional duty if the crews were not broken up and dispersed to other duties as is usually the case. The availability of rested crews made it possible to rotate home the deployed crews of the (very similar) ships that could then remain on station.

The second Sea Swap exchange involved identifying pairs of Arleigh Burke-class guided missile destroyers and scheduling each of those pairs to swap crews. This would allow the fleet commander to keep one of the ships on station for 11.5 months instead of the more usual 5 months resulting from subtracting transit time from a 6-month deployment. The idea was to wring nearly a year out of the deployed time of two crews and ships, rather than the more usual 10 months on station realized from two ships and crews deploying in the more usual way.

If the Navy needs to rotate crews among ships its leaders will be able to do so in a more informed way as a result of Sea Swap. As is sometimes the case in pilot programs most of the issues that would arise during the program were identified in advance. Even so, some of the issues proved to be more important than expected and some less. In both cases subsequent use of ships and crews will benefit in a variety of ways by virtue of the Navy's leadership having taken advantage of the opportunity to try something new and look carefully at the results.

[16]U.S. Government Accountability Office. 2004. *Force Structure: Navy Needs to Fully Evaluate Options and Provide Standard Guidance for Implementing Surface Ship Rotational Crewing,* GAO-05-10, Washington, D.C., November 10.

SUGGESTIONS FOR EXPERIMENTATION

Putting some of the recommendations of the committee into practice on a trial basis can provide the Navy and Marine Corps with the necessary information to determine their impact on transformation. Integrating the insights and lessons derived from experimentation into policy changes will enhance the prospects for successful innovation and transformation.

Career Paths: Some Experimentation

The Embedded Oval—Pilot Test

The Navy's plans for manning ships in the future will require a level of multitasking that may or may not be realistic, given issues related to recruitment, assignment, and training. This experiment is designed to simulate anticipated littoral combat ship (LCS) requirements in a familiar and controlled environment.

The LCS will be manned by a new crew. Planning dictates that the ship's crew will be much smaller, more senior, and more experienced than that of other ships of this size and capability.

Historically the shape of the chart that portrays the numbers of sailors at each pay grade in Navy activities at sea and ashore has resembled a pyramid. Many positions are at the bottom of the pyramid, and there are fewer at each succeeding pay grade above. Many seamen, firemen, and airmen are overseen by fewer petty officers, and those by still fewer chiefs and junior officers, in turn responsible to a single commanding officer.

For the LCS the shape of the equivalent personnel structure has been described instead as an oval. The Navy's manpower and personnel system, and its training and learning establishment in particular, will have to adjust to the needs of a ship which demands the experience that a sailor gains as a junior member of a ship's crew but provides none of that experience.

The Navy expects that in the near future the lack of junior positions and people in the crew will not be a problem. The Navy will tailor the commissioning crews of the first few of these special ships carefully, assigning sailors from the large, traditional career specialties to new training also tailored for these ships. In the long run, though, the Navy must fold all of the new ships' recruiting, training, manning, advancement, and career requirements into its routine business. It is not clear to the committee that the Navy has determined at this point what all of the implications are of this new class of ship, whose numbers may grow quickly.

The committee suggests a trial to help Navy leaders evaluate their plans for dealing with the special requirements imposed by the crewing needs of the LCS and other new ship classes. At least six new issues arising from the Navy's LCS crewing concept appear worthy of testing:

1. Sailors on the LCS will be asked to learn multiple skills and perform in multiple roles. Can Navy schools teach "hybrid sailors" and traditional specialists simultaneously? How will sailors perform in operational environments when tasked heavily in these multiple roles?

2. Sailors will have to train to qualify. The sailor production process is to promise that new arrivals will not need much, if any, additional time to learn on the job. There is to be little learning curve for new crew members. No time for that is built in to the manning plan for the ship.

3. Billets aboard the LCS will be for journeymen and masters. Apprentice preparations for future LCS crew members will have to occur elsewhere in the personnel system. In steady state where will that be?

4. Sailors who have served on the LCS will have different experiences and skills from those of their brothers and sisters who have served elsewhere at sea. Will LCS sailors be unsuitable for routine assignments in their specialties? How can their special competencies be sustained to be reused in subsequent LCS assignments?

5. Sailors on the LCS (and increasingly on other new ship classes) will be "one deep" with no backups. How will the distribution systems for officers and enlisted personnel manage to achieve perfection in manning the ships, providing the right person, every time, all the time?

6. LCS crew members will possess skills and aptitudes that are much in demand both in the Navy and in the civilian world. Will LCS crewmembers' pay be adequate for the new skill sets? How will the promotion process need to be adjusted to recognize their new training, skills, and careers?

Here is a way that new procedures and concepts for accommodating LCS could be tested on a larger scale at the same time as the Navy is moving forward in actually manning the first LCS crews. The committee suggests that subsets of existing combatant crews assigned to existing ships be used to simulate the crews of LCSs in service, as follows:

1. Select the 40 billets in the manning document of DDG-51-class ships that most resemble the billets to be manned in the basic crew of the LCS. Call this the "embedded DDG/LCS." Under current plans the shape that will best represent the grade structure of the LCS is an oval. The committee calls this the "embedded oval."

2. Determine the degree to which the requirements of these DDG billets differ from their LCS counterparts.

3. Rewrite the billet descriptions in the embedded DDG/LCS oval to eliminate the difference, and adjust the descriptions of the associated DDG work centers to ensure that the work of the DDG host still gets done. Some duties might move from one work center to another. The idea would be to adapt the host DDG

to supporting the embedded LCS, and to employ the embedded LCS crew members in ways that approximate their LCS duties as closely as possible.

4. Train the embedded DDG/LCS incumbents and inbound replacements to the new DDG/LCS billet requirements.

5. Manage the sailors who leave the embedded DDG/LCS billets in the way that Navy leaders plan for all LCS-experienced sailors to be managed eventually.

Smaller crews will demand nearly perfect matching of numbers and skills to requirements. The Navy's personnel system can test its intended processes for achieving near perfection in LCS by using the same processes for precisely supporting the embedded DDG/LCS. The crew's experience on the LCS would be valuable in normal service because future Navy ships will require skill sets similar to what the LCS would require.

Since the embedded DDG/LCS crew members will be performing in an operational ship well before LCSs join the fleet in numbers, the Navy Personnel Research, Studies, and Technology Office of the Navy Personnel Command will have something like an operational laboratory in which to test the performance of hybrid sailors (both officer and enlisted) under somewhat realistic circumstances.

The committee understands that the Navy is working on plans to support the special roles that men and women will play in LCSs, as well as on processes to ensure high levels of crew productivity. Forming these embedded DDG/LCS test beds in ships in operating environments, training the people for them, stressing the crew, assigning them to follow-on tours, and keeping their numbers just right will reveal a good deal about the efficacy of those plans. The committee believes that this type of test will reduce the eventual risk of manning an LCS without appreciable risk to the operating fleet today.

On/Off Ramp—A Proposal for a Controlled Randomized Experiment

Professionals in industry and academia are given the opportunity to take a sabbatical leave for the purpose of fulfilling a specific personal goal (e.g., writing a book, traveling extensively for research, and taking family leave for starting a family or caring for an ailing family member). In software and systems development companies, law firms, consulting agencies, and other high-stress industries, sabbaticals have been recognized as a useful management tool in countering burnout. It is estimated that 14 to 24 percent of corporations in America have established sabbatical programs.

Today 37 percent of all professional women leave the workforce at some point in their careers. Most say they want to come back, but the challenges of finding work after having been out of the workforce for some period of time mean that only a fraction finally do—and few of them work full-time jobs.[17]

The Navy and Marine Corps are also faced with the prospect that many skilled sailors and marines (officers and enlisted) whose talents are highly valued by the naval forces may face burnout or choose to leave the service to attain more schooling, develop new foreign language skills, or tend to family issues.

Consideration of sabbaticals and other types of temporary leave options is not new to the Department of Defense. In 2003 the RAND Corporation conducted an analysis for the DOD on extended leaves for officers.[18] Descriptive in nature, the work lays out various kinds of extended leave programs, evaluates the return on investment likely from different programs, and offers recommendations for specific programs as well as observations about how extended leave programs, more broadly, might be instituted. One could expand on this idea with an experiment to test the assumptions of the value of various leave options to the Navy.

To begin the research some preliminary focus group and survey data collection activities can be conducted to gauge the interest service members might have in various leave offerings. This would be followed by the design of an experiment such as the one described below, to help the Navy explore the notion of the on/off ramp—where people can put their obligation (if any) on hold, take a leave from the Navy (or Marine Corps) for a predetermined period, and return to the service and continue their active duty.

An outline of the experiment follows. Consistent with best experimentation practices, the committee recommends that a multidisciplinary team of advisers (as described above) be assembled to work out the detailed design.

1. The option for a sabbatical would be offered to a randomly selected group of sailors (or marines) in both the enlisted and officer ranks.

2. Only specific job skills in hard-to-fill, high-demand fields based on rating, Military Operation Specialty, and Navy Enlisted Code would be eligible for the experiment.

3. The sailor's or marine's time in service would pause when he or she is put on sabbatical and renewed upon return to active duty (i.e., the clock stops while the service member is pursuing another endeavor).

4. The size of the experiment would have to be sufficient to meet the experimental objectives to produce reliable estimates for the predetermined metrics.

5. The duration of the experiment would be approximately four years, long enough to measure outcomes to determine or at least estimate the success of the approach and any possible unintended consequences.

---

[17]Tamara J. Erickson, Robert Morrison, and Ken Dychtwald. 2006. *Workforce Crisis: How to Beat the Coming Shortage of Skills and Talent*, Harvard Business School Press, Boston, April.

[18]Harry J. Thie, Margaret C. Harrell, and Marc Thibault. 2003. *Officer Sabbaticals: Analysis of Extended Leave Options,* RAND Corporation, Santa Monica, Calif.

TABLE 4.1 Treatment and Controls for On/Off Ramp Experiment

Sailor or Marine Status	Treatment	Control (not eligible for the experiment)
Officer—still has an obligation	Clock stops on their obligation, and resumes upon return to active duty	Same rank and designator, matched for age and other demographics
Officer—beyond initial service obligation	Clock stops on their obligation, and resumes upon return to active duty	Same rank and designator, matched for age and other demographics
Enlisted E-5	Clock stops on their obligation, and resumes upon return to active duty	Same Navy enlisted codes, matched for age and other demographics

6. The analysis plan would be set up to measure specific outcomes, including retention behavior, achievements and accomplishments attained when off the ramp, and enhanced value to the Navy.

It is entirely possible that sabbaticals, and the career pause that accompanies them, will permit many career professionals to do on their own exactly what their services would like them to do anyway. Examples of sabbatical pursuits that would benefit the Navy and Marine Corps include language training and regional travel; advanced degrees using GI Bill benefits; creating more stable family situations to avoid future difficulties; making a religious or philanthropic contribution to society.

Some of the parameters are described in Table 4.1.

Lateral Entry Pilot Demonstration

The Navy and Marine Corps appreciate the value of cross-fertilization of ideas and experiences with other organizations, as demonstrated over the years by provision of a variety of personnel exchange programs, opportunities for schooling for marines and sailors, experience tours, and short programs like "Executives to Sea." The committee believes that creating a more fundamental kind of exchange of people, skills, and perspectives has a place in the manpower and personnel system for a transformed naval force.

At least four new issues or opportunities present themselves for the transformed force in the new, distributed battle space: (1) the Navy will need sailors with advanced knowledge and experience in ships that will provide no billet base in which sailors can gain seniority and that experience (LCS); (2) the Navy and Marine Corps increasingly will need specific skills that heretofore have been scarce in the naval services but are important to the conduct of all phases of the war (especially phase IV)[19] and are now in short supply among uniformed personnel; (3) skills and experiences the services value can be taught by others; thus sailors and marines can join the services with high-level skills achieved elsewhere, for which the services do not have to pay; (4) the nation will benefit from more interaction between ordinary citizens and their military, and especially service members; in particular, policy makers and those who influence the career choices of America's young people need more exposure to the value of military experience and training. It is possible that all of these issues can be addressed to some degree through a greatly expanded program of lateral entry and exit for sailors and marines.

The committee recommends that both the Navy and Marine Corps undertake a pilot demonstration, a test of limited and focused accession of sailors and marines in several areas to determine the degree to which such a program could address the issues above and to assess the willingness of people from the private sector to join at an advanced pay grade for a specified period. It is the committee's view that civilians at an advanced stage of their careers would potentially be attracted to a period of Navy service for several reasons. First, there are the advantages the Navy advertises to potential new recruits: travel, the opportunity to serve their country, the chance to gain solid work experience. The committee believes it is possible that people in the midst of careers in the civilian sector would often see a period of military service, and the opportunities offered in that service, as valuable additions to their resumes.

The lateral entry and exit demonstration could explore at the same time some of the alternatives for the sabbatical experiment suggested earlier. In a group offered the opportunity for a sabbatical, for example, those who did not choose to leave the service to take advantage of a brief separation opportunity could choose a sabbatical to experience another professional area of military interest.

The naval services often need skills and experience that are not fully developed in their uniformed members. The services sometimes need skills only temporarily or have an emerging requirement that demands skills that are out of the ordinary, and there are a variety of potential sources of such expertise (e.g., contractors, nondefense government personnel, academia). It is possible that people with the skills the services need would want to serve in the Navy or Marine Corps on a limited basis, yet be unwilling to serve for an entire career or even to accept the full-length obligation that regularly accompanies an enlistment or commissioning. The services could benefit from people such as these serving for some period of time on active duty, and such people would ultimately return to their prior employment with genuine Navy or Marine Corps experience. The committee believes that these people would function as service advocates. During the period of temporary service, the individual's command would benefit from his

[19]Phase IV of a campaign is activities conducted after decisive combat operations to stabilize and reconstruct the area of operations.

or her functioning as an integral member of the command and having the added credibility of being a member of the team in full uniformed status.

The committee recommends that a pilot program be undertaken to evaluate the availability of such talent, the willingness of people with such talent to serve in this capacity, and the effective contribution that such people could bring to a real-world naval service situation. There are numerous current situations that could be considered for such a pilot. For instance, the lead ship of the LCS class could benefit from having as a member of the crew (in uniform) one or more members of the ship's or ship system's design or build team. Such a person would be of immense help to the crew as it worked through the initial operations. The expert's service could continue through the initial deployment of the ship. The individual(s) involved would return to their civilian jobs imbued with increased real-world experience. Their civilian employers would undoubtedly see these returning crew members as more valuable employees with improved perspective on the operational and personnel environment in which Navy ships perform.

The standup and operation of Joint Task Force Horn of Africa or Africa Command could benefit from having resident experts with cultural knowledge and language skills as uniformed experts embedded in each command. People with such skills could be obtained from other governmental agencies such as the Department of State and the U.S. Agency for International Development or from academia. Having this expertise on the command staff would make initial policy formulation and execution much more accurate and effective. These new commands could also benefit from the service of an individual with organizational design experience who could be involved on a daily basis as the command is stood up to ensure that the command has from the outset all the advantages of being designed for effectiveness and responsiveness. Such expertise also could be obtained from academia or from organizational design expert contractors who would don uniforms. Upon completion of their assignments, these individuals would return to their civilian jobs with the benefit of understanding how the naval services function and be in position to be a more fully informed asset to their employer, as well as the services.

The transformation of the naval service manpower system could benefit from having real corporate world and academic expert assistance in the areas of organizational design, systems engineering, and human relations design and better enabling implementation of a human capital strategy. Although this might be seen as able to be accomplished using reservists, existing active-duty personnel or contract personnel, those avenues have shortcomings. The committee believes that some individuals who possess the appropriate skills and would welcome the opportunity to serve for a brief period might be wary of the more open-ended commitment of reserve service, particularly in the wake of recent stop-loss orders and IRR call-ups. Moreover, using recognized or certified assistance from outside the reserve or active-duty forces could bring fresh perspectives, experience, and input. Such individuals would "live the life," gaining organizational perspective and at the same time credibility with team members as someone other than an outsider. Once the term of service was completed, both parties would have gained from the experience.

Before starting such a pilot demonstration project, specific criteria for measurement of impact and success must be developed. Criteria should begin with project identification and proceed through recruitment, training, assignment, work assignment, and definition of outcomes ending in final pilot evaluation. No project should be commenced without a complete set of criteria against which clear success or failure can be determined.

Such a pilot study would require some special permissions or authorities to execute. There would need to be a provision for enlistment or commissioning such individuals for limited time frames. Agreements with employers may need to be negotiated. Just-in-time training may need to be delivered to ensure full ability to integrate into the military command.

Compensation: A Pilot Program in Sea Pay

Chapter 3 discussed various potential changes to the services' retirement and compensation systems. The committee understands the need for congressional, or at least DOD, authorities to undertake many of the possible reforms that the committee addressed. Still, the Navy and Marine Corps have the authority needed to conduct some trials or pilots in compensation. Special pays in particular constitute an area where the services have substantial latitude to try new things.

The Navy and Marine Corps pay career sea pay in accordance with a formula that considers the grade of the recipient and the amount of sea duty served. The services could examine paying career sea pay in a different way. Having succeeded in using assignment incentive pay (AIP) to encourage qualified sailors to volunteer for duty in somewhat isolated locations using an auction system, the Navy could lead the way in examining possible sea pay changes by considering using an auction system for a trial group of sailors for that pay too. It is recognized that there is no shortage of creativity on the Navy's part in encouraging sailors to undertake the more difficult duties essential for operations at sea. Recently the Navy instituted a pilot program aimed at testing the effects of offering to compensate sailors specifically for extending tours at sea or shortening tours ashore to go to sea.[20] These two versions, called Sea Duty Incentive Compensation, Extend (SDIC-E) or Curtail (SDIC-C), offer sailors in selected ratings (ratings with shortages at sea) $500 to $750 per month for the months of adjustment of their sea duty tours.

The committee applauds this initiative and agrees that this offer will improve

---

[20]Chief of Naval Personnel Public Affairs. 2007. "Pilot Program to Support a More Sea Centric Navy," Bureau of Naval Personnel message, March 20, Washington, D.C. See <http://www.npc.navy.mil/AboutUs/NPC/PublicAffairs/NewsDetails/SeaCentric.htm>. Accessed on October 5, 2007.

manning at sea in a focused way. It believes that a pilot program would be useful in helping the Navy to determine the appropriate size of the incentive by exploring the "take rate" for the incentives of various sizes.

While SDIC addresses specific shortages at sea, the committee questions whether the payment of career sea pay in blanket amounts based on pay grade and months at sea is the most efficient use of the resources. There may be a better way to determine the value of the incentive if the Navy's intent is to use sea pay to offset each individual's perceived personal and professional cost of going to sea. The Navy could build on its experience with both AIP and SDIC to test a sea pay auction approach to the filling of sea duty billets in general.

The Navy employs an excellent tool for conducting assignment transactions among detailers (assignment officials) and enlisted sailors. This system, called Super-JASS (Job Assignment Selection System), permits important information to flow between detailer and constituent and allows enlisted sailors to make informed decisions about their assignments and careers. It is used already to negotiate the auction prices of positions in the AIP. It could be used also to create an experiment in the allocation of sailors to sea duty by variable economic incentive that might replace the blanket amount of sea pay now in use.

Encouraging Warfare Behaviors and Service Appropriate to the Requirements of Irregular Warfare

In the committee's view, the naval services need to recognize deliberately the service and encourage the behaviors that are appropriate and necessary to the transformed naval service in the new reality of warfare. Some important changes are underway and others are needed to permit the services to create and retain the special capabilities they now require in their people.

The Navy and Marine Corps are encouraging sailors and marines to acquire a broad range of skills needed in the expanded missions of the global war on terror and irregular warfare. Events of the last 4½ years have shown, however, that certain knowledge and skills are in short supply in the naval services. Chapter 2 discusses the importance of foreign language ability, regional experience, being able to contribute to nationbuilding (construction, policing, provisionally governing, accomplishing other civil affairs tasks), and cultural and ethnic awareness. The Cold War demanded much less agility from the naval services on these dimensions than it appears will be required in coming years with the new naval missions.

The Navy and Marine Corps have focused in the past on combat arms and warfighting skills. This approach has served the military needs of the nation well. Specialties in combat support and combat service support have been sustained largely through partitioning them off from competition with the combat arms for advancement of specialists and allocation of resources within the services.

There is a pertinent and instructive Navy example of the problem of creating an important specialty for officers that was not walled off from competing with those serving in warfare specialties. Navy leaders have made more than one attempt over the years to create cadres of officers with regional expertise and language ability, officers who retained their warfare specialty designations at the same time. In the 1980s the Navy's Country, Area, Regional Specialist (CARS) and Country, Area, Regional Specialty Officer (CARSO) programs were created and sailors, both officer and senior enlisted, were encouraged to become specialists in one or another part of the world. Developing and sustaining area and language specialties, however, distracted warriors (or so it seemed to promotion and advancement selection boards) from their primary warfare duties. Without high-level sponsorship and the commitment of the Navy to sustaining their skills, members of the community gradually lost out in competition with warfare specialists for promotion. The CARS and CARSO officer programs survived for little more than a decade, although the need remained. Based on this and other experiences of the services in building specialty communities, the committee believes that the Navy and Marine Corps will need a better plan for equipping sailors and marines with the skills they need to be both warriors and builders of nations and relationships for coming missions.

Many sailors and most marines have contributed in significant ways to the difficult efforts required in these early years of the global war on terror. Specific recognition of their contributions could be a powerful incentive for others, to ensure that they contribute also. Reviews of administrative data, for instance, by boards convened to conduct the reviews, could provide specific details on marines' and sailors' contributions. Having that information, the CMC and the CNO might choose to publicize the extent of these contributions. This could have the effect on the one hand of praising those who have done so much, while on the other hand illustrating which skills have been called on again and again in the new operational environment.

The services could consider some tangible recognition of those who achieve important added skills and put those skills to use: Individuals who strive for sound leadership and other combat skills, while at the same time learning what it takes to communicate and build relationships in the new environments and missions, would be recognized in special ways. The committee believes that the services should focus on the outcome when determining the changes needed for the manpower and personnel systems for the transformed force.

During the course of this study, many briefings included a discussion of the impact of both external and internal forces that influence the manpower and personnel decisions and challenges that the Navy and Marine Corps face. Experiments, surveys, analyses, and pilot demonstrations inform these decisions by collecting data that will estimate the potential impacts of proposed policy changes. Therefore the committee offers the following finding and recommendation:

Finding: Establishing and sustaining transformation in Navy and Marine Corps manpower and personnel policies will require the use of a variety of research tools.

Research tools ranging from surveys and analyses of administrative data to pilot demonstrations and experimentation have always been key tools in formulating policies that support military manpower transformation. Experimentation has been one of the research tools that the Department of Defense and the Navy in particular have used in achieving changes to the shape of military forces, and the Navy's recent Sea Swap project illustrates that the Navy understands the value and challenges of experimentation.

Recommendation: The Assistant Secretary of the Navy for Manpower and Reserve Affairs, the Chief of Naval Personnel, and the Deputy Commandant of the Marine Corps for Manpower and Reserve Affairs should continue to develop and use research tools and experimentation to address the many facets of the naval services' transformation equation. In particular, the committee recommends the following projects:

- Evaluate Navy plans for dealing with the special requirements imposed by the crewing needs of the littoral combat ship (LCS) and other ship classes by using "embedded ovals," that is, subsets of existing combatant crews, assigned to existing ships, to simulate the crews of LCSs in service.
- Design and conduct a controlled experiment to test the assumptions of the value of various leave options, including the notion of the on/off ramp.
- Undertake a pilot demonstration of a lateral entry and exit program to evaluate the availability of needed talent outside the services, the willingness of people with such talent to serve for a limited duration, and the contribution such people could bring to real-world naval service situations.
- Use the Navy's Job Assignment Selection System, Super-JASS, to create an experiment in the allocation of sailors to sea duty by using a variable economic incentive that could replace the blanket amount of sea pay now in use.
- Use administrative data and other research tools to identify warfare behaviors and service appropriate to the requirements of irregular warfare, and encourage such behaviors and service by publicizing and rewarding them.

5

Transforming the Naval Force: Obstacles and Strategies for Implementation

Our naval forces must continually change to develop capabilities to address current and emerging missions and contingencies. In recent years the Department of the Navy has become a high-velocity enterprise—the situations to which it must respond change very rapidly and will continue to do so for the foreseeable future. This change requires the Department of the Navy to establish human resource policies that are sufficiently agile to promote—rather than impede—the naval services' required responsiveness.

The committee has concluded that many current policies, which were adopted during an era when agility requirements were not nearly as great as they are today, are inadequate. The committee also concluded that the needed changes include some solutions that are under the direct control of the Department of the Navy, some solutions that cross organizational lines (are joint), and some solutions that require regulatory and budget support from Congress. In many instances solutions that might be under direct Department of the Navy control cannot be addressed in isolation because of linkages to other concerns that are *not* under the Department of the Navy's control. This chapter explores several types of impediments to change in the area of naval manpower and personnel and suggests ways of overcoming them.

There are many potential obstacles to organizational change and reform. Surmounting those barriers can be particularly challenging within an organization such as the naval services, in which culture, traditions, hierarchy, and bureaucratic procedures play important roles. Although organizational change has received immense attention from scholars and practitioners alike—in the mid-1990s Van de Ven and Poole[1] reported a count of a million articles on the topic—outcomes depend on so many attributes of the particular organization, context, and reforms being undertaken that no set method or process can guarantee successful implementation of the desired change.

Any decision maker reading this report scarcely needs input from this committee to appreciate the challenges associated with changing large-scale organizations, particularly in the domain of manpower and personnel concerns within the military. In fact, instances of fundamental change in the policies surrounding U.S. military personnel since 1947 are rare. Two major exceptions are the end of the draft in 1973 and the broad changes in officer training, career paths, and requirements for promotion to flag or general officer ranks that were mandated by the Goldwater-Nichols Act of 1986.[2] In both cases fundamental change was adopted in a matter of a few years, and in both cases the changes have endured the test of time.

Four factors stand out as reasons for the rapid and enduring adoption of both the all-volunteer force (AVF) and the Goldwater-Nichols changes. The first is a sense of crisis. The AVF was adopted as the war in Vietnam ended, public support for the military was at a low point, and the U.S. Army was in disarray. The Goldwater-Nichols Act ensued in the wake of the humiliating failure of the joint effort in Operation Desert One of 1980, in which U.S. forces attempted to rescue hostages from Iran.

The second factor that stands out is the importance of sustained, top-down leadership. In the shift to the AVF, President Richard Nixon and Secretary of Defense Melvin Laird embraced the change and laid the groundwork for it. Secretary Laird worked closely with Congress, established an internal committee to push the change forward, and held the services accountable for implementing the many changes required. The Goldwater-Nichols Act is an example of outside leadership: congressional proponents, including Senators Barry Goldwater and Sam Nunn and Representatives Les Aspin and Bill Nichols, engaged vigorously in the change effort over several years.

A third critical factor in both cases was the establishment of an external commission or oversight group. The Gates Commission on an All-Volunteer Armed

[1]Sergio Fernandez and Hal G. Rainey. 2006. "Managing Successful Organizational Change in the Public Sector," *Public Administrative Review,* March/April, p. 168 (Andrew H. Van de Ven and M. Scott Poole. 1995. "Explaining Development and Change in Organizations," *Academy of Management Review, Vol.* 20, No. 3, pp. 510-540).

[2]For a discussion of the change process leading to adoption of the all-volunteer force, see Bernard D. Rostker and Curtis L. Gilroy, 2007, "The U.S. Experience," *Service to Country: Personnel Policy and the Transformation of Western Militaries,* Curtis L. Gilroy and Cindy Williams (eds.), MIT Press, Cambridge, Mass., pp. 233-262. For the Goldwater-Nichols case as a rare instance of successful personnel reform, see Arnold L. Punaro, 2004, "Leadership and Perseverance," *Filling the Ranks: Transforming the U.S. Military Personnel System,* Cindy Williams (ed.), MIT Press, Cambridge, Mass., pp. 282-283.

Force, established by President Nixon on March 27, 1969, was instrumental in building support for and shaping the decisions related to the AVF. A major external study led by Center for Strategic and International Studies, a Washington think tank, was crucial in the Goldwater-Nichols case.

Both the AVF and the Goldwater-Nichols changes owe much of their success and durability to the research and evaluation that surrounded their adoption. In the AVF case detailed, multidisciplinary research pulled together by the Gates Commision was crucial in building support for change and also for establishing key details of implementation. Research and evaluation continue to be instrumental. Support for and specifics of the Goldwater-Nichols Act relied heavily on studies commissioned and conducted by the House and Senate Armed Services Committees.

These four factors and others described in this chapter will be important in building and sustaining support for fundamental changes in manpower and personnel policies in the naval services in the future. The committee's conversations with military leaders and review of written materials indicate that Department of the Navy leadership has a keen appreciation of the challenges involved in transformational change, the merits of various reforms and strategies for implementing them, and how both vary across contexts. By "strategies," the committee means preparing the Navy for change using the best current theories and practices of organizational change, searching the path ahead for obstacles, and preparing plans to overcome or circumvent those obstacles.

In formulating its recommendations and considering strategies for implementing them, the committee was mindful that it possesses limited information about numerous organizational, political, and cultural factors that matter greatly in decisions about what manpower and personnel reforms should be undertaken and how. Accordingly, it focused much of its attention on trying to provide templates or frameworks to assist informed decision makers in pursuing transformation and weighing the trade-offs involved, rather than presuming that it has the right answers to the questions of what should be done, when, and how.

The committee believes that the success of any organizational change depends on attention up front to two interrelated concerns: (1) the nature and magnitude of the organizational obstacles that must be surmounted and (2) the sequencing of change and what is demanded at each stage of the change.

The next section discusses each of these in general, as a prelude to considering how they bear on implementing the specific recommendations adopted by the committee. The chapter goes on to identify several categories of obstacles to change. It then matches several broad categories of manpower and personnel reform to those categories of obstacles. The result is an illustrative mapping of the transformation landscape for naval manpower and personnel policy. The chapter ends with findings and recommendations aimed at improving the effectiveness of the Department of the Navy in managing and implementing such changes.

STAGES OF ORGANIZATIONAL CHANGE

In identifying potential sources of resistance and means of overcoming them, it is useful to distinguish five stages in the organizational change process:

1. Preparing the organization for change,
2. Conceptualizing and planning the change,
3. Implementation,
4. Evaluation, and
5. Institutionalizing the change.

The value of planning for each of these stages, in a coordinated fashion, would appear self-evident, yet both the scholarly literature on organizational change and the experience of practitioners highlight how seldom this is done effectively. Each of these stages in general is briefly discussed below. Later in this chapter the committee offers specific suggestions for managing each of the stages in the context of specific recommended reforms.

Preparing the Organization for Change

The importance of this stage cannot be overstated. As noted above, diverse stakeholders often have competing interests with regard to specific manpower and personnel policies. Moreover, tradition plays a central and invaluable role in military institutions, so attachment to past practices is to be expected. Articulating a compelling case for change entails highlighting what is deficient about the status quo (and how it will be rectified) or what would be superior about the different future that is being sought through the change initiative.

The following paragraphs describe a direct approach to preparing the organization for change. Although aware of the effectiveness of the indirect approach in some instances,[3] the committee concluded that the current situation called for a more timely and direct change. The committee was persuaded, based on the briefings it received, that the Navy needs to move quickly and decisively and does not have the time for the indirect approach.

Creating a compelling case for change will require that the Chief of Naval Operations (CNO) and the Commandant of the Marine Corps (CMC) be personally involved in describing and rationalizing important personnel changes. If they are not directly and visibly involved the change process will be devalued by many service people and constituencies. In creating a compelling up-front case for change, influential "champions" must be identified and provided with appropriate resources and support. Potential winners and losers with organizational influence

[3]See Stephen Peter Rosen, 2004, "Implementing Changes in U.S. Military Personnel Policy," *Filling the Ranks: Transforming the U.S. Military Personnel System,* Cindy Williams (ed.), MIT Press, Cambridge, Mass., pp. 297-299.

should be identified. The overall goal must be to discover and implement those changes that are best for the organization. Nevertheless, losers will oppose the changes. Communication and influence strategies should be developed for influential groups who are likely to perceive themselves as losers from the change efforts. For example, current members of the force serve with the expectation that current pension and health care arrangements will continue. These programs are expensive, and the Department of the Navy is under significant pressure to reduce cost. Because efforts to reduce labor costs will likely diminish benefits for some members already in service, they will face opposition from service members and retirees. Therefore, there will likely need to be some adjustments, such as grandfathering or lump-sum payouts to gain the support of those who will lose.

Winners, on the other hand, can be co-opted to establish constituencies with a vested interest in seeing change efforts prevail, even in the face of potential resistance and drift. For example, single members stand to benefit from the elimination of the "with dependents" pay differential. Communication and influence strategies should be developed for influential groups who are likely to perceive themselves as either winners or losers from the change efforts.

External threats, crises, or transformations can weaken resistance to change and sometimes provide a compelling rationale for transformation. The committee believes that a substantial downsizing of U.S. forces in Iraq may provide a valuable opportunity to reframe future military needs and priorities and launch an agenda for change in manpower and personnel policies within the Navy and Marine Corps.

Healthy competition—among or within service branches—can also be a powerful catalyst for innovation. An effective way to prepare an organization for transformation is by causing individuals to perceive change initiatives as providing engaging opportunities to earn esteem (and possibly other rewards) for themselves and the unit to which they are attached.

Legendary General Electric Chief Executive Officer Jack Welch introduced into the management lexicon the phrase "relentless and boring" to describe what is required of those leading organizational and cultural change. Few of us aspire to be relentless and boring—and few of us believe this is what our superiors will reward. Consequently it is easy for leaders to lose interest in sustaining the persistent, focused communications that are essential to paving the way for organizational change.

Conceptualizing and Planning the Change

Given the complex, complementary, and challenging nature of changes in manpower and personnel policy, responsibility for conceiving and planning policy changes should be coordinated centrally to the greatest extent possible, even if this approach requires working around existing organizational distinctions within the service branches. Participation builds commitment, so involving major constituencies whose buy-in will be essential for change initiatives to succeed is highly desirable, even if this slows somewhat the pace at which change occurs.

Numerous factors exacerbate the challenge of implementing organizational change in the Department of the Navy. Some of the more obvious ones are political and budgetary processes, frequent changes in military leadership, necessary commitments to legacy assets, and rapid environmental and technological shifts, all of which can hamper sustained efforts at programmatic change. These obstacles are understood all too well by the various parties whose cooperation or resistance will determine the success of the change initiatives. In conceptualizing and planning change in this context, leaders must articulate a compelling and credible plan for overcoming the major impediments to change.

Implementation

A number of the recommendations offered in this report point to the need to articulate an overarching human capital strategy (HCS) for the Department of the Navy. As presented, all of the ideas contained in the HCS seem reasonable and well intentioned, but it is highly unlikely that all can be accomplished simultaneously or within anticipated resource constraints. It is necessary, therefore, that the naval services work to put primary focus where primary focus is needed, not simply to document an open-ended wish list of actions that could be taken under ideal circumstances.

A completed strategy makes possible coordinated programmatic change initiatives that will prove more effective over the long haul than a plethora of "one-off" interventions addressing specific individual concerns. The committee urges naval leaders to give serious consideration to launching strategically designed demonstration projects that can assess the interdependent effects of the changes in technology, work design, staffing levels, and human resource management practices subsumed under the rubric of "transformation." Surveys and analyses, pilot demonstrations, and experiments of the sort described in Chapter 4 can be useful tools for anticipating how the effects of specific proposed changes are likely to vary across organizational contexts.

Case studies of large-scale change efforts suggest several practices that may enhance success. Sequencing implementation to ensure early and vivid initial victories and capturing of "low hanging fruit" generally builds support for subsequent phases. Buy-in among line managers is usually enhanced by allowing local customization of broad programmatic objectives to meet the needs of managers across diverse organizational circumstances.

The process of conceiving and implementing change sends powerful signals about what an organization truly values and how it seeks to function going forward. If flexibility, responsibility, and teamwork are values being elevated, then a change process that is itself highly centralized, is inflexible, and reinforces individual (rather than group) accountability undermines the objectives it seeks.

Care should be taken to structure implementation processes in ways that strongly reinforce the cultural messages that manpower and personnel policies are intended to convey.

Evaluation

Rigorous evaluation will be an essential input into reform and transformation efforts—not only rigorous ongoing and retrospective assessments of initiatives that have already been implemented but also prospective assessments of various policy alternatives. Whatever the course of future transformation, the committee believes that increased flexibility along numerous dimensions will be required of naval personnel (and of the Department of the Navy as an organization). The need for enhanced flexibility is likely to be manifested in more varied assignments and mastery of more varied skills at any point in time; more variety among individuals in the sequencing of assignments and development of careers over time; more variety in sources of labor; possibly more varied means by which individuals establish and maintain relationships with the military; and more variety in roles, routines, practices, and policies across organizational subunits. Surveys and analyses of administrative data, pilot demonstrations, experimentation, and other forms of evaluation research can play an invaluable role in formulating means of achieving greater flexibility along these dimensions, as well as in assessing the relative merits of particular initiatives that have been undertaken. The committee has outlined several worthwhile areas and methods of inquiry in Chapter 4. Evaluation research might also profitably focus on the development of metrics by which decision makers can gauge the effectiveness of manpower and personnel policies on an ongoing basis.

Institutionalizing the Change

In a context where capital assets are as costly and long-lived as in the military, there is an inherent danger that, as introduced in Chapter 1, human capital becomes the cushion by which those who allocate budgetary resources adjust to the constraints they face. If transformation and the global war on terror (GWOT) truly entail fundamental changes in concepts of operations, technology, and organizations, then significant long-term investments in human capital and organizational development will be required. Unfortunately if the case for transformation must be made on an annual basis, little fundamental change is likely to occur. Careful attention must be devoted up front to marshalling formal mechanisms (e.g., metrics, incentives, systems, processes) and informal means (e.g., culture, routines and rituals, and communication practices) to institutionalize new ways of managing the Department of the Navy's human resources. Put differently, a comprehensive HCS to support transformation must be built to last.

OBSTACLES TO CHANGE

Any attempt to categorize impediments to organizational change is likely to be somewhat artificial. Nonetheless, during its deliberations the committee found that categorizing various types of obstacles was helpful in determining the order for undertaking specific initiatives as well as the kinds of catalysts and implementation tactics that might facilitate particular changes. The following paragraphs summarize four categories of obstacles to organizational transformation. Later in this chapter the committee employs this taxonomy to suggest a framework for prioritizing reform initiatives and developing implementation strategies.

Informational Obstacles

The first category of constraints involves the availability of information needed to make and evaluate decisions, as well as the organizational capability for acquiring and analyzing that information. The need for clear and credible communication about strategic change is imperative. Organizational change becomes more challenging when an organization has greater need for (and less internal capacity to produce) information in four arenas:

1. *Generating consensus for change.* A logical first step in removing some barriers to change and reform is to clarify to all stakeholders why a specific change is necessary. The rationale for the change must be clearly stated and understood by all. It is not enough to say that we need to implement this change because things just are not working. Evidence of how and why things are not working must be provided, as well as evidence that some new approach is capable of providing superior outcomes. When neither the status quo nor the future is viewed as problematic, or when alternative approaches are not viewed as meritorious, the challenge of organizational change increases dramatically.

2. *Analysis of relative merits of alternatives.* Most problems have numerous solutions. Possible solutions must be vetted and analyzed, with both benefits and liabilities thoroughly and credibly examined. The proposed alternative must be clearly stated, understandable, and able to withstand rigorous scrutiny. For example, consider a proposal for preventive programs aimed at reducing stress during lengthy deployments. It might be relatively straightforward to assess the outcomes and costs of such a program relative to results associated with treating stress after the fact.

3. *Analysis of trade-offs or opportunity costs.* When outcomes are longer-term, more uncertain, more multifaceted, more difficult to assess, and more costly, it is easier for detractors to derail change initiatives by arguing that resources would be better deployed elsewhere. Consider a proposal to adopt 360-degree

feedback as part of a new approach to career development.[4] It is difficult not only to quantify the benefits and costs of such an initiative, but also to assert definitively that the resources required to implement and sustain this reform will yield a higher return on investment than other strategies would yield. Specific proposals may entail costs that are very difficult to quantify, such as second- or third-order effects of making a change (or not doing so) on personnel morale, retention, equipment use, maintenance cycles, and the like. The costs of delaying, only partially implementing, or not implementing a particular change may be significant but difficult to quantify in a timely and persuasive manner.

4. *Assessment of change impact.* For the same reasons discussed above, the impact of some proposed interventions cannot be assessed as credibly, objectively, promptly, or inexpensively as others. Mobilizing and sustaining change will be significantly more challenging in such circumstances. Note that problems in assessing the effect of changes can arise from many sources. Consider an initiative that is likely to span various leadership regimes. If leadership support is integral to success (e.g., implementing 360-degree feedback would require significant leadership support), it will be difficult to determine whether a perceived failure is attributable to the intervention itself or to the change in leadership regime. Knowing this up front, decision makers may be reticent to undertake such initiatives, relative to ones that generate faster and more straightforward outcome measures.

Organizational (Department of the Navy) Obstacles

Various attributes of the naval service are likely to exacerbate the challenges associated with organizational change. The committee focuses on three of these:

1. *Culture.* Naval forces are steeped in tradition, and culture plays a vital part in the ethos of the Navy and Marine Corps. U.S. naval forces have been successful in armed conflict throughout our nation's history, and the organizations and their members remain strongly attached to their heritage. Culture is a positive influence in defining our naval forces but can be a hindrance to the introduction and acceptance of changes.

2. *Winners versus losers.* Most organizational changes have distributional consequences—sometimes dramatic ones. When the group of those who perceive themselves to be losers from a proposed initiative are more numerous, vocal, widespread, and influential, the costs of implementing change increase precipitously.

[4]A 360-degree feedback or multi-rater feedback is a technique for evaluating an individual's performance that is based on input from the target individual, the boss, and subordinates. Information may be collected via surveys or face-to-face interviews. Results are typically used for improving individual performance, increasing individual accountability, enhancing team performance, furthering organizational change, and providing evidence for human resource actions.

3. *Collateral impacts.* The naval forces are complex organizations, and specific changes frequently cause unanticipated results in nontargeted areas. Prior to implementation, changes must be examined for both near- and long-term, second- and third-order effects on the force as a whole. Even so, surprises should be expected.

Timing Obstacles

Features of the timing or implementation of particular reforms can materially affect the obstacles to change.

1. *Future commitments.* Changes that entail future commitments can have unanticipated effects and therefore must be thoroughly examined from all aspects prior to implementation. The costs of such changes can be hard to estimate accurately beforehand. In an organizational context characterized by high turnover among those who lead and those who allocate resources (including DOD and White House civilians as well as Congress), making credible future commitments is nontrivial. This is particularly so if those who are affected in the short term will not experience the anticipated future benefits.

2. *Grandfathering.* In the same vein, when a proposal may reduce benefits or limit opportunity for vested or current service members, the change will be opposed by those affected and their support groups on the grounds that an explicit or implicit contract is being abrogated. Changes of this nature generally require an adjustment (e.g., grandfathering, lump-sum payment) to gain the support of those who would otherwise lose out.

3. *Ability to change incrementally.* Changes that can be implemented incrementally (e.g., along specific time lines, events, budget years, performance standards) are often easier to implement and sustain than changes that cause sudden or large actions. Systems must be in place to capture support data for sustained incremental change. Compare, for instance, efforts to change the retirement system with changes to recruitment practices. It is much harder to envision a phased implementation of a fundamental retirement overhaul than a phased change in recruitment policies.

4. *Ability to adapt.* Organizations, their members, and particular reforms will vary in the ease with which feedback can be incorporated into the change process. Some interventions are by their very nature nearly impossible to undo. Change will be easier to implement and sustain in settings where feedback loops with quantifiable data exist and are attended to.

External Obstacles

A host of factors outside the immediate control and purview of the Department of the Navy obviously can impede or facilitate change. The committee focuses on four sets of influences:

1. *Joint service.* Manpower and personnel policy changes implemented within the naval service will affect other services and may be met with resistance from them. Such changes require considerably more attention to obstacles that may arise elsewhere within the military establishment, within Congress, and in the arena of public opinion.

2. *Legal and regulatory.* Some changes are beyond the Department of the Navy's implementation authority. Fundamental changes to the retirement system as recommended in Chapter 3 will require changes to the Defense Officer Personnel Management Act (DOPMA) and Reserve Officer Personnel Management Act (ROPMA) provisions of Title 10. Depending on how the reform handles the additional front-loading of compensation, a retirement overhaul may also require changing the pay table, which is set in law. Any change that would increase the fraction of senior officers will require changes to DOPMA and possibly ROPMA. Changing the pay table to a time-in-grade structure will also require congressional approval.

3. *Fiscal.* In some cases, budgetary realities and competition for funding are large obstacles to change. In other cases, they may be the stimulus to change. DOD plans call for budgets to grow less rapidly over the coming 6 years than they did between 1998 and 2007. The costs of equipment and manpower are on the rise, however, raising deep concerns about the future affordability of today's force. Unless they will save money proposed changes will be subject to rigorous scrutiny during budget formulation. Thus the fiscal impact of proposed changes, including the cost of not changing, must be examined.

4. *Public perception.* Proposed changes will vary in the extent to which their merits will be closely scrutinized by currently serving sailors and marines and by other constituencies, including retirees, contractors, Congress, lobbies, the media, and the public at large. Such scrutiny can be an important factor in determining the magnitude and pace of change in manpower and personnel practices within the armed services.

MAPPING THE TRANSFORMATION LANDSCAPE: MATCHING THE OBSTACLES TO REFORM PROPOSALS

Based on discussions with Navy and Marine Corps leaders and reviewing Department of the Navy materials regarding existing manpower and personnel policies and objectives, the committee identified a roster of domains and topics that are likely to be involved in the evolution of a transformed naval force:

1. Completing and articulating a comprehensive human resources strategy
2. Domains of manpower and personnel practice reform
    - 2.1. Cash compensation
    - 2.2. Training and development
    - 2.3. Career paths

   2.4. Job design (e.g., skills, multitasking, and so on)
   2.5. Recruitment and staffing
   2.6. Performance evaluation
   2.7. Performance assessment
   2.8. Retirement system
   2.9. Health care benefits
3. Critical crosscutting topics
   3.1. Stress and morale
   3.2. Quality of family life (e.g., bases, housing)
   3.3. Automation and technology impacts
   3.4. Sourcing (e.g., role of reservists, civilian employees, contractors)

The committee believes that Item 1 is an essential prerequisite to making progress in the other areas of reform and transformation. Item 2 enumerates areas of human resource practice that will be profoundly implicated in efforts to transform the naval force. Item 3 lists several manpower and personnel topic areas that the committee believes will be of crucial importance to the Department of the Navy and will require comprehensive study and solutions that draw on most or all of the human resource domains listed under Item 2.

Cross-referencing these domains and topics against the various types of obstacles to organizational change furnishes a useful heuristic. The resulting matrix, shown in Table 5.1, has several useful functions. Identifying the sources and magnitude of resistance in changing certain human resource practices can illuminate key priorities at each of the five stages of the change process described earlier in this chapter. For instance, when obstacles are primarily *informational*, more up-front effort will clearly be required in making the case and preparing the organization for change. Experimental or pilot studies may be crucial for evaluating initial change initiatives, implying a more incremental implementation process in order to build acceptance.

When major obstacles are *organizational* (within the Department of the Navy), much greater attention must be devoted to identifying external developments and internal champions that can be leveraged in building a credible case for change. The burden on senior leadership to be "relentless and boring" in their communications, central to any large-scale organizational change effort, becomes even greater. Understanding who the likely proponents and opponents will be—and how formal and informal connections among individuals and groups are likely to influence the development of support or dissent—is an invaluable resource in determining how to plan and implement the change, whom to coopt, and where initial implementation efforts might begin to build momentum.

Reforms facing significant *timing obstacles* necessitate more attention on how proposed changes will be sustained and institutionalized, particularly when those responsible for leading the change and allocating the requisite resources have short tenures. Identifying stable internal and external constituents whose contin-

TABLE 5.1 Gauging Obstacles to Manpower and Personnel Change in the Naval Services: An Illustrative Analysis

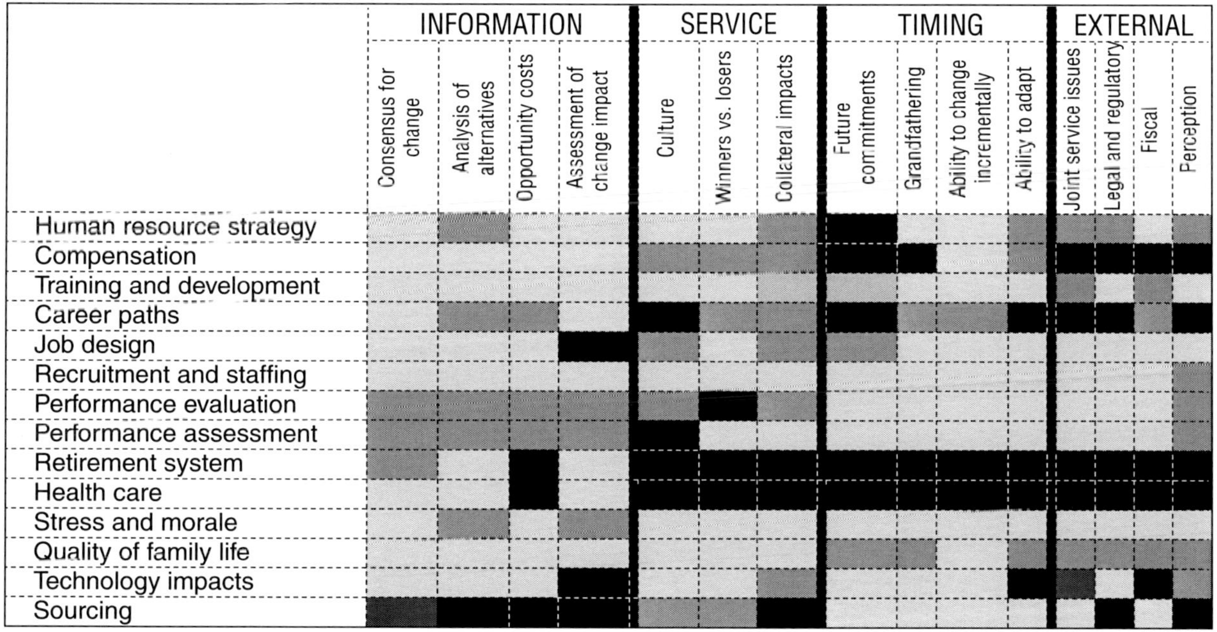

Obstacles that can feasibly be managed within the Department of the Navy without significant changes in standard practices, routines, and procedures.

Obstacles likely to require more concerted effort and coordination (e.g., realignments, task forces, new roles) and internal championing.

Obstacles that are unlikely to be surmounted without external pressures from external champions to impel and sustain long-term change and foster accountability.

ued backing can be marshaled, and involving them early in the planning process to secure commitment, are critical. Early victories may be crucial to secure ongoing support. This suggests focusing implementation and evaluation strategies on areas likely to produce rapid, measurable positive results.

Manpower and personnel reforms confronting *external obstacles* demand early alliances with key gatekeepers whose support or opposition is decisive. In the context of the Department of the Navy this will in many cases involve enlisting the support of other service branches, congressional units, lobbying groups representing military constituencies, the media, and the public. Surmounting external obstacles frequently requires first building strong and vocal consensus internally, so that external opponents cannot exploit conflicts inside the organization in seeking to derail reforms.

A typical shortcoming among organizations seeking to enlist allies in surmounting external obstacles is the failure to institutionalize the alliances early in the change process. Senior leaders among various service branches, for instance, may agree on the need for a specific personnel reform that would benefit all of them. However, unless strong consensus is established early on among those whose cooperation will be necessary to mobilize external support and implement the proposed changes, progress may come to a halt as those leaders turn their attention to other pressing concerns.

A heuristic like Table 5.1 can also provide useful summary insights into the unique challenges of transforming particular human resource domains by virtue of the internal and external constraints likely to be encountered. Obviously, the nature and magnitude of resistance in any area (such as compensation) will depend greatly on the precise nature of the reforms being sought and on local circumstances. To illustrate how the Navy Department might use this heuristic in formulating transformation strategies, the committee attempted to fill in the matrix and analyze the resulting patterns.

Within each of the human resource topic areas listed in the rows of Table 5.1, the committee considered the range of feasible reforms that decision makers might pursue—from incremental to transformational—and assessed worst-case scenarios for the obstacles likely to be encountered. Working across each row, the committee sought to determine what source(s) of resistance were likely to be most acute, in relative terms, within a specific domain of reform. Working down the columns, the committee assessed what human resource domains were likely to be most strongly affected by a potential source of resistance.

As a shorthand device in Table 5.1, the committee coded the challenges associated with organizational change into three broad categories:

1. Those that it believes can feasibly be managed within the Department of the Navy without significant changes in standard practices, routines, and procedures (light gray);

2. Those that it believes will require more concerted effort and coordination (e.g., realignments, task forces, new roles) and internal championing (gray); and

3. Those that it believes are unlikely to be surmounted without pressures from external champions to impel and sustain long-term change and foster accountability (black).

There is no accepted and precise methodology for making such judgments. The committee's assessments—subjective, to be sure—reflected four sources of information: (1) presentations to the committee from department officials regarding previous efforts to change manpower and personnel policies, (2) experience with Department of the Navy manpower and personnel policies among members serving on the committee, (3) previous studies bearing on particular Department of the Navy manpower or personnel policies within and efforts to change them, and (4) research findings regarding organizational change processes in the literature. For instance, organizational scholars have identified particular organizational features and environmental contexts in which inertial forces are most intense,[5] and the committee incorporated such findings into its appraisals of the implementation challenges likely to be associated with various domains of human resource transformation.

Table 5.1 summarizes the committee's assessments, with light gray, gray, and black shading being used to depict challenges that it classified as falling into categories 1, 2, and 3, respectively.

A graphical representation of the information in Table 5.1 offers some insights into the transformational landscape confronting the naval services. Figure 5.1 plots the various human resource domains and topics in Table 5.1.[6] The dotted lines in the figure show the median values for the external and internal constraint dimensions, breaking the figure into four quadrants. The figure highlights several distinct clusters of human resource domains based on the likely obstacles to be encountered in transformation.

In the lower left quadrant of Figure 5.1 is a set of topic areas that, in the committee's view, are amenable to reform efforts that can feasibly be managed within the Department of the Navy without significant changes in standard practices, routines, and procedures. This cluster includes articulation of a completed and comprehensive human resource strategy, recruitment, programs addressing stress and morale, and training and development initiatives. In the top right quad-

---

[5]See Michael T. Hannan and John H. Freeman, 1984, "Structural Inertia and Organizational Change," *American Sociological Review*, Vol. 49, pp. 149-164; and William P. Barnett and Glenn R. Carroll, 1995, "Modeling Internal Organizational Change," *Annual Review of Sociology*, John Hagan (ed.), Vol. 21, pp. 217-236, Annual Reviews, Inc., Palo Alto, Calif.

[6]In quantifying the constraints in Figure 5.1, the committee assigned values of 1, 2, and 4 to the light gray, gray, and black cells of Table 5.1, respectively. Accordingly, in Figure 5.1, the total score for internal constraints can range from 7 (least intense) to 28 (most intense); for external constraints, the range of possible values is 4 to 16.

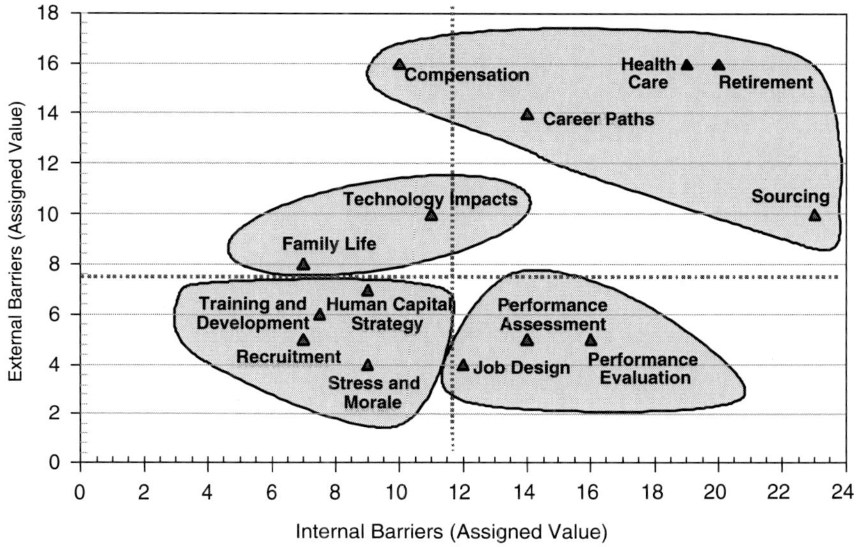

FIGURE 5.1 The transformational landscape depicting human resource domains based on the likely obstacles to be encountered.

rant of Figure 5.1 are topic areas that confront more intense potential obstacles both internally and externally: retirement, health care, privatization and sourcing, career paths, and (to a lesser degree) cash compensation. In the committee's judgment surmounting resistance and entrenchment in these areas will require help from external champions to impel and sustain long-term change and foster accountability.

The topics falling in the lower right quadrant and (to a lesser extent) in the upper left quadrant involve sizable implementation challenges due to the likelihood of major obstacles either internally or externally. The committee's estimation is that reform and transformation in these areas—job design, performance evaluation and assessment, and programs addressing the quality of military family life and the effects of automation and technology on military personnel—will likely require more concerted effort and coordination (e.g., realignments, task forces, new roles) and internal championing.

Table 5.1 reflects the committee's collective judgments, and below in this chapter the committee describes broadly the types of internal and external championing efforts envisioned to support transformation in arenas confronting intense obstacles internally and externally. But one hastens to reiterate that these investigations are meant to be *illustrative* and *exploratory*. Future assessments should incorporate systematic methods for dealing with differences among individuals' judgments concerning the organizational challenges associated with specific manpower and personnel reforms. The committee finds that even such simple heuristics can be useful in working through the organizational implications and implementation challenges associated with broad human resource strategies. The potential value in this exercise does not come from accepting the committee's judgments about the nature and magnitude of implementation challenges for particular areas of reform. Rather, the value will be realized if policy makers are able to formulate their own judgments and use a tool like this to organize how they think, plan, and act with regard to transformation.

Potential Obstacles to Implementing the Committee's Recommendations

In Chapters 2 and 3 the committee offers recommendations to naval leaders in several of the generic topic areas discussed in the previous section. This section looks at a few of the committee's recommendations through the lens of Figure 5.1 and discusses specific hurdles that they may encounter.

Of the recommendations offered by this committee, those related to cash compensation and the retirement system probably involve the most serious external impediments. For example, changing the pay table to one based on time in grade rather than time in service would require congressional approval. Moreover, such a change would affect the other services. Thus as described in the previous section, such pay table reform would meet both substantial external resistance and some internal resistance.

As discussed in Chapter 3, a fundamental overhaul of the retirement system like the three-part reform recommended by the committee would raise both internal and external obstacles. In 2006 Congress paved the way for the longer careers that such an overhaul would encourage in some career fields, by changing the pay table to reflect up to 40 years of service. Remaining obstacles include opposition from retiree associations; changes that would be needed in legal authorities, including the retirement, DOPMA and ROPMA, and "up-or-out" provisions of Title 10; concerns over internal equity between those under the existing system versus the new system; and establishment of the base of information that will be needed to set the particulars of the new system. For example, policy makers will need to determine which occupations should be selected for the encouragement of longer or shorter careers. The specific amounts of gate pays or other retention pays will also need to be determined. Overcoming such obstacles will require a comprehensive implementation strategy.

Fundamental changes to the health care system for military families and retirees could raise similar obstacles. On the other hand, the less ambitious reforms recommended by this committee—requiring retirees under age 65 to stipulate a primary and secondary insurer and providing them with an incentive to choose Tricare as the secondary insurer when they have other options, or indexing Tricare to the annual cost-of-living adjustment for the military retirement annuity—should entail far less internal resistance. Of course, providing incentives for retirees to use their civilian-employer-provided plans would shift costs from the government to those civilian employers. To the extent that powerful employers are involved and lobby vigorously against the shift, Congress may be reluctant to allow the change.

In contrast, many of the recommendations of this committee involve actions that are largely under the Department of the Navy's control. For example:

- *Expansion of Naval Junior Reserve Officer Training Corps (NJROTC).* Expanding this program would have a relatively small effect on funding and manpower requirements. Expansion of NJROTC could face external resistance from schools, communities, or parents, however.
- *Development of better evaluation programs to measure and interpret changes in workload, separation from home, length of repetition of deployments, and similar factors.* This recommendation can be implemented by the Department of the Navy without involving external players. Because the other services face similar issues, however, it would make good sense for the Department of the Navy to work together with the other military departments and with the Under Secretary of Defense for Personnel and Readiness on programs to evaluate shared issues.
- *Allowing officers to remain in operational or specialist billets, with limitations on rank, for their entire careers.* Naval leaders on their own can sponsor the study to investigate how to embed performance incentives, as recommended by the committee. As discussed earlier, however, changes to the basic pay table will require support from the other services and changes in law. Independent of pay table considerations, making a reformed system work well would probably require changing the "up-or-out" provisions and possibly the DOPMA and ROPMA rules on the distribution of officer ranks that are elements of Title 10. Changing career progression in this way might also raise substantial internal opposition, because the new operational or specialist careers would not be directed toward traditional executive positions.
- *Enlisted-to-commissioned promotion path.* Implementing this recommendation will probably require a major cultural change within the Navy and Marine Corps. Shifting individuals directly from the senior enlisted ranks into field-grade officer status can raise social and cultural issues about the differences between officers and enlisted personnel that are difficult for the services to look squarely in the eye.
- *Lateral entry.* The services already use lateral entry for some positions, for example in the medical professions. More widespread use of lateral entry may raise cultural resistance within the Department of the Navy. Lateral entry into leadership roles would run up against important traditions and expectations related to command and hierarchy. On the other hand, lateral entry for individuals with new skill sets (information technology specialists) at midlevel (E 4-5 or O 2-4) might face the least opposition.

- *Career on/off-ramps and sabbaticals.* Implementing a system that would allow for easier on/off-ramps involves obstacles that must be addressed directly by Department of the Navy policy. Ensuring fairness and holding at no harm the careers of individuals who choose this path will be essential. The promotion system and design of career progression may be the most difficult of these. The extent of the external obstacles will be determined in large part by the design of the system with regard to length of breaks, status of member during the breaks, and benefits and compensation policy during the breaks. The most difficult of these obstacles may be refining the compensation and retirement systems to make allowances for these policies.

Sabbaticals will raise issues within the Department of the Navy and also from external agents. External obstacles related to retirement credit, continuation of health care benefits, and cash compensation while on sabbatical may require changes in legislative authorities. Department of the Navy obstacles related to the effect of a sabbatical on a service member's career path and performance evaluation require assessment. Sabbaticals for foreign area officer or high-technology training opportunities might face the least opposition.

FINDINGS AND RECOMMENDATIONS

Comprehensive Human Resource Strategy

In 2004 the Department of the Navy published its human capital strategy (HCS)[7] outlining the department's broad goals for human resources. Two years later, in 2006, the Navy *Strategic Vision: MPT&E [Manpower, Personnel, Training, and Education] Strategic Plan*[8] was published as the first of three milestones leading to implementation of the HCS. Later that year the second milestone, the Navy MPT&E *One Voice Reference Book*,[9] was put forth. As of this writing the third milestone—the Navy MPT&E roadmaps—containing implementation action plans, discrete tasks required to achieve the strategic vision, and metrics, accountability methods, and timelines for completion are yet to be finished. Thus the implementation of the HCS effort remains unfinished.

---

[7]William A. Navas Jr., Assistant Secretary of the Navy, Manpower and Reserve Affairs; LtGen Garry L. Parks, USMC, Deputy Commandant, Manpower and Reserve Affairs; and VADM Gerald L. Hoewing, USN, Chief of Naval Personnel. 2004. *Department of the Navy Human Capital Strategy,* Department of the Navy, Washington, D.C., June 21.

[8]Navy Manpower, Personnel, Training, and Education (VADM J.C. Harvey Jr., USN, Deputy Chief of Naval Operations for Manpower, Personnel, Training, and Education, N1). 2006. *Strategic Vision: MPT&E Strategic Plan,* Department of the Navy, Washington, D.C., October 31.

[9]Navy Manpower, Personnel, Training, and Education (VADM J.C. Harvey Jr., USN, Deputy Chief of Naval Operations for Manpower, Personnel, Training, and Education, N1). 2006. *One Voice Reference Book,* Department of the Navy, Washington, D.C., November 1.

Finding: Completing, communicating, and implementing a comprehensive human capital strategy will be essential to achieving the Navy and Marine Corps transformation goals.

In 2004 the Department of the Navy published an overview document outlining an HCS. The committee is concerned that the strategy remains incomplete, in that the declared third step, the implementing roadmaps, have yet to be issued by the Navy. Nor is the Department of the Navy HCS broadly understood or routinely employed as a guide to decision making outside a relatively narrow circle of manpower and personnel specialists. Moreover, since the signing of the Navy's HCS in June 2004, the services have undertaken important operational and organizational changes. They have also gained experience related to the Navy's downsizing, lengthy commitments in Iraq and Afghanistan, and increased reliance on the private sector—all without the coherence and logic that a completed HCS would bring to the decision-making process.

Completing and communicating an HCS is an essential prerequisite for achieving success in more specific initiatives relating to transformation of the Department of the Navy. For reasons articulated in Chapter 1, investments in any of the domains or topic areas in Table 5.1 will reap greater returns—and be easier to justify internally and externally—when undertaken under the aegis of a comprehensive and coordinated strategy for transformation, rather than on a piecemeal basis. As documented in Figure 5.1, articulating such a strategy involves relatively few internal and external obstacles and modest commitments of resources. In short, the committee believes that this initiative involves relatively little investment, few roadblocks, and the potential for extraordinarily high return.

To illustrate the desiderata of a comprehensive human resource strategy, consider the case of Southwest Airlines.[10] The company pursues a niche strategy, offering short-haul, point-to-point, low-cost flights between underutilized airports. It seeks to compete based on cost, reliability, convenience, and superior customer service. Given the high fixed costs of an airline (as is the case for the Department of the Navy), efficient utilization of capital is a critical determinant of profitability. By minimizing delays and turnaround times, Southwest can offer more flights per day on a given route, providing a higher return on invested capital and greater convenience to its target customers. If Southwest employees are able to inject fun, humor, and enthusiasm into the travel experience, customers are likely to be less disturbed by some practices the airline uses to enhance operational efficiency (e.g., no reserved seats, no meals). Thus, successful execution of the airline's strategy requires a workforce that exhibits high productivity (high output at relatively low cost), a strong service orientation, flexibility (e.g., in tasks, shifts), teamwork, enthusiasm, and stability. The company views its human resource practices as dif-

[10]Jody Hoffer Gittel. 2002. *The Southwest Airlines Way: Using the Power of Relationships to Achieve High Performance*, McGraw-Hill, New York.

ferent channels by which it markets or brands itself to internal customers, insisting that congruent and consistent messages are sent through these various channels. Indeed, it hires people with marketing backgrounds into the human resources function. (Historically, the head of human resources at Southwest and the head of customer relations have reported to the same senior executive, on the grounds that they are both engaged in branding and serving customers—employees and passengers, respectively.)

The result of this clear strategy and explicit branding is a set of human resource practices that powerfully complement one another and explicitly support the key success factors. Recruitment, selection, compensation, training, performance management, and other human resource practices reinforce the importance of teamwork, flexibility, service, and productivity. Rank-and-file employees can articulate the business strategy, the key success factors, and how the company's culture and human resource practices support the firm's mission.[11]

Recommendation: The Chief of Naval Operations (CNO) and the Commandant of the Marine Corps (CMC) should take ownership of their services' human capital strategy (HCS) and direct its prompt completion. Beyond that, the CNO and the CMC should institute a process to review and update their HCS in light of changes in the strategic environment, future plans, and evolving experience with existing human resource policies. The completion of the services' HCS should be done with the following criteria in mind:

1. *Aligned.* The HCS should be linked clearly to the services' goals and missions, identifying the highest-priority "key success factors" required of personnel for organizational success.
2. *Internalized.* The HCS must be communicated to and broadly understood at all levels, in ways that clarify to individuals in each subunit how their efforts affect overall success.
3. *Routinized.* The HCS should routinely inform decisions, trade-offs, and resource allocations and should be embedded in everyday operating procedures (e.g., planning and budgeting, personnel reviews, external reporting).
4. *Coherent.* The HCS should promote coherence and synergies in human resource administration across specific domains (e.g., recruitment, compensation, training, and development). It should sustain a human resources "brand" that makes clear to current and prospective sailors and marines what is expected of them and what they in turn can expect of the organization.
5. *Measurable.* The HCS should describe desired outcomes that can be and are assessed with metrics.
6. *Adaptable.* The HCS should be dynamic, undergoing routine reassessment

[11] Libby Sartain and Mark Schumann. 2006. *Brand from the Inside: Eight Essentials to Emotionally Connect Your Employees to Your Business*, Jossey-Bass, San Francisco.

and adjustment in light of learning and of changing organizational and environmental contingencies.

7. *Consequential.* Supporting the HCS should represent (and must be *perceived* to represent) a significant element in the formal assessment and evaluation of leaders.

The committee recommends an assessment of current proposed human resource strategies against this list of criteria and creation of a template simple enough in form and content that it can be used to articulate the key success factors and human resource strategy to diverse audiences at all levels of the naval services.

Propelling Transformation in Human Resource Practices

Finding: The implementation of changes to manpower and personnel policies is a complex and difficult process.

The SECNAV, CNO, and CMC already have the authorities they need to make some changes. Even in those cases, however, change may be complex or involve several stakeholders, including the individuals who serve, their families, military retirees, the other military services, and Congress. If manpower and personnel policies are going to meet the challenges of the future, the Department of the Navy must prepare for and overcome strong opposition.

Recommendation: To bolster the likelihood that changes to manpower and personnel policies will be adopted, the Secretary of the Navy, working with the Chief of Naval Operations and the Commandant of the Marine Corps, should take control of the change management processes. The Department of the Navy-wide processes should include the following:

- Opportunities for early successes,
- A program of continuing evaluation, and
- The institutionalization of a program of comprehensive education of and communication with service members and other stakeholders about the reasons for the changes to manpower and personnel policies and the desired outcomes of change.

It is noted that some areas of human resource transformation confront internal and external constraints of sufficient magnitude to require creating unique entities—inside and outside the Department of the Navy—that can champion and coordinate change efforts. How such entities are constituted and tasked will of course depend on the particular transformation strategy and agenda adopted, but several generic suggestions are offered.

Internal Task Force-like Function

Some changes recommended in this report can be accomplished within the existing authorities and responsibilities of the Department of the Navy, but will entail extraordinary complexity or cut across the claimancies of one or more stakeholders within the department. These changes may require the establishment of a task force or task force-like organization internal to the department to develop and implement change(s). In these cases such an organization should be stood up under the auspices of the CNO's or CMC's office and given the necessary resources, authorities, and responsibilities to design the desired outcomes and resolve conflicts between the stakeholders to successfully complete a change. Examples of such ad hoc organizations that have worked well for the Navy are the Executive Review of Naval Training and Task Force Excel—both constituted to design and execute the revolution in Navy training.

External Oversight Group

Particular domains—retirement, health care, sourcing, career paths, and compensation—face internal and external obstacles of sufficient magnitude and complexity that significant organizational change is unlikely to be achieved without external pressures to impel and sustain long-term progress and foster accountability. There are numerous examples within the military and the corporate sector of change initiatives that have been supported by such an external forcing factor. The Defense Advisory Committee on Women in the Service (DACOWITS), for instance, adopted this approach. The Base Realignment and Closure (BRAC) process took yet another step and placed the initiating step in the hands of an external group.

Within corporate America, companies like Deloitte and Touche and Ben & Jerry's have adopted similar approaches in pursuing objectives relating to workplace equity, corporate social responsibility, and environmental sustainability. Such an oversight group is crucial not only for pursuing transformation in areas facing internal and external obstacles but also for ensuring that reforms in other manpower and personnel domains are implemented in a coherent and coordinated manner that supports the Department of the Navy HCS.

The specific structure, function, and composition of such an external oversight body would of course depend on the specific transformational objectives of naval leadership and its assessment of what changes are feasible within a given environment. Nonetheless, the committee offers the following general guidelines concerning the formation and functioning of such an oversight entity:

1. The group should be relatively small, containing members who collectively bring to their work the following competencies:

   1.1 Distinguished military experience, stellar credibility, and extensive personal connections within the naval services and to other services and branches of the federal government;

1.2 Distinguished corporate leadership experience, particularly in the areas of human resource management and organizational change;

1.3 Expertise in the legal, regulatory, and fiscal constraints governing manpower and personnel policies within the Navy, Marine Corps, and other service branches;

1.4 Access to the most recent findings from the scholarly literature and management practice for the topic areas under the group's purview;

1.5 Knowledge of manpower and personnel policies—as well as initiatives being undertaken—in other service branches;

1.6 Access to accurate and timely information regarding the characteristics, needs, and preferences of:

1.6.1 Former, current, and prospective service members (active, reserve, and civilian) and their family members;

1.6.2 Organizations representing specific interest groups and constituencies;

1.6.3 Contractor organizations;

1.7 No personal or professional stake in the outcome other than public service; and

1.8 Expertise in communications, public relations, and influence strategies.

2. Stability of group membership is needed, given the long planning and implementation cycles involved in transforming manpower and personnel policies.

3. Also needed is unfettered access to information, personnel, and the media aside from limitations required by national security considerations.

4. The group should be responsible for producing on a regular basis candid, specific, and public assessments of progress toward achieving the transformational objectives articulated by naval leadership.

5. The group should be responsible for providing regular briefings to the Secretary of Defense, the Secretary of the Navy, the Chief of Naval Operations, the Commandant of the Marine Corps, and representatives from the Senate and House Armed Services Committees, assessing progress to date and remaining challenges and opportunities relating to the transformed naval force.

CONCLUDING STATEMENTS

The committee began its task by noting the current superiority of the U.S. naval forces and the fact that at the heart of that greatness are the men and women who serve in uniform. With this fact in mind and guidance from the CNO, the committee projected likely future circumstances, including future combat scenarios; likely scientific and technological developments; and the characteristics of the recruitment age U.S. population. The committee's analysis was guided throughout by the manpower and personnel issues outlined in the terms of reference for the study. As requested by the CNO and to the best of its ability, the committee has developed several recommendations presented throughout the report in the context of pertinent analysis and discussion. In addition, the major recommendations are presented in the Summary at the outset of the report.

The committee shares the sense of urgency expressed by the CNO that the Navy develop leaders for the 21st century and by the CMC that the Marine Corps continue to develop the individual marine as the heart and soul of the organization. The committee sincerely hopes that its efforts will be of help to them as they lead their services into the future, and it looks forward to observing as the Department of the Navy implements the changes in manpower and personnel policy necessary to maintain its superiority.

Appendixes

A

Committee and Staff Biographies

John H. Moxley III (IOM) is the retired managing director of the North American Health Care Division, Korn/Ferry International. His areas of expertise include training, costs, and manpower issues; federal government agency administration; military medical issues; and chemical, biological, radiological, and nuclear detection. He has held a number of senior positions in academia, government, and industry, including dean of both the University of Maryland and the University of California, San Diego, Medical Schools; assistant secretary of defense for health affairs; and senior vice president at American Medical International. Dr. Moxley has served on numerous scientific boards and advisory committees, including the American Hospital Association Board of Trustees, California Medical Association, American Medical Association, National Fund for Medical Education, and the Henry M. Jackson Foundation for the Advancement of Military Medicine. He was chair of the National Research Council (NRC) Committee on Strategies to Protect the Health of Deployed U.S. Forces and served on the NRC Board on Army Science and Technology. Dr. Moxley is a member of the Institute of Medicine and the Naval Studies Board.

Cindy Williams is a principal research scientist of the Security Studies Program at the Massachusetts Institute of Technology. Her areas of expertise include national security budgetary issues and reform of military personnel and pay policies. She has served as assistant director for national security at the Congressional Budget Office, where she led the National Security Division in studies of budgetary and policy choices related to defense and international security. She has also served as a director and in other capacities at the MITRE Corporation in Bedford, Massachusetts; as a member of the senior executive service in the Office of the Secretary of Defense at the Pentagon; and as a mathematician at the RAND Corporation in Santa Monica, California. She is the editor and one of several authors of *Filling the Ranks: Transforming the U.S. Military Personnel System*, a book published in 2004 about military personnel policy choices for the future and the coeditor of *Service to Country:Personnel Policy and the Transformation of Western Militaries*, a book published in 2007 about military personnel policy choices in North America and Europe. Dr. Williams is a member of the Naval Studies Board.

Beth J. Asch is a senior economist at the National Security Research Institute at the RAND Corporation. Her areas of expertise include personnel economics and defense manpower. Her research focuses on empirical analyses of recruiting and retention policies and estimating models of effects of alternative pay and retirement systems on recruiting, retention, ability sorting, and productivity of personnel in the armed services. She recently served on the staff of the Defense Advisory Committee on Military Compensation and has served as a member of the working group of the seventh and the ninth Quadrennial Reviews of Military Compensation. She also served as a visiting faculty member at the University of California, Los Angeles, Department of Economics and as an associate economist at the CNA Corporation. Dr. Asch is a current faculty member at the Pardee RAND Graduate School.

James N. Baron is the William S. Beinecke Professor of Management at the Yale School of Management. His areas of expertise include human resources; organizational design and behavior; social stratification and inequality; work, labor markets, and careers; and economic sociology. He codirected the Stanford Project on Emerging Companies, a large-scale study of the organizational design, human resource management practices, and financial and nonfinancial performance of entrepreneurial firms in Silicon Valley. Dr. Baron served as a member of the NRC Committee on Strategic Education Research Plan: Bridging Research and Practice and the NRC Committee on Performance Appraisal.

Owen R. Cote Jr. is associate director of the MIT Security Studies Program. He has a broad base of expertise, as both a political scientist and a technologist. He has served as assistant director of the International Security Program at Harvard University's Belfer Center for Science and International Affairs. He is coeditor of the journal *International Security*. His most recent book is *The Third Battle of the Atlantic,* a history of how the U.S. Navy performed antisubmarine warfare during the Cold War. Dr. Cote is presently working on a book that compares competing theories of the sources of innovation in military doctrine.

Lee F. Gunn retired from the U.S. Navy with the rank of vice admiral after 35 years of service and is currently president of the Institute of Public Research at the Center for Naval Analyses Corporation. His background is in surface warfare as well as in manpower, personnel, and training issues. He commanded the USS *Barbey* (FF 1088), Destroyer Squadron 31, and Amphibious Group 3. He has served as deputy chief of naval personnel, and commander, Navy Personnel Command, during which time he played a key role in redesigning the Navy's manpower and personnel establishment. In particular, he instituted the Navy's operational cost management training and evaluation program, and reconfigured the Navy's major command management inspection process to teach and evaluate cost management. Admiral Gunn also served as inspector general of the Department of the Navy and was responsible for the department's overall inspection program and its assessments of readiness, training, and quality of service.

James L. Herdt retired from the U.S. Navy with the rank of master chief petty officer of the Navy and is currently chief executive officer and president of Herdt Consulting, Inc. His consulting firm specializes in human resources management, organizational design and behavior, change management, and institutional learning and training. His background is in naval operations both ashore and afloat, including human resources management and training. In his last position he served as the ninth master chief petty officer of the Navy (the senior-most enlisted person in the Navy) where his responsibilities included leading program and policy development for the enlisted force.

Barry M. Horowitz (NAE) is professor of systems engineering at the University of Virginia. His areas of expertise include the design and development of large-scale networks and information systems; application of security technology to large, network-based commercial systems; and design of large systems that involve coupling private data systems or mission critical support systems with open networks, such as the Internet. He has served as chairman and founder of Concept Five Technologies and as president and chief executive officer of the MITRE Corporation and of Mitretek Systems. He has served on numerous scientific boards and advisory committees, including as a member of the NRC Committee on C4ISR for Future Naval Strike Groups, and he is a new member of the Naval Studies Board. Dr. Horowitz is also a member of the National Academy of Engineering.

Leon A. Johnson retired from the U.S. Air Force Reserves with the rank of brigadier general after 33 years of service and is currently a manager for United Parcel Service (UPS) flight operations. He has served as the employment manager, employee services manager, chief pilot, employee relations manager, and training manager during his 17 years with UPS flight operations. During his Air Force career, he served as the mobilization assistant to the Assistant Secretary of the Air Force for Manpower and Reserve Affairs. In that role he advised senior Air Force leadership on outreach, marketing, retention, and recruiting initiatives. He also served as the chair of the Air Force Reserve Command (AFRC) Human Resources

Development Council (HRDC). As the chair of the HRDC General Johnson was the principal staff officer responsible for formulating and administering, in concert with other Air Force Reserve Staff agencies, policies and programs for and affecting AFRC people programs, including outreach and retention initiatives. General Johnson is a member of several organizations, including the Air Force Association, Reserve Officers Association, League of United Latin American Citizens, Women in Aviation, and the International Black Aerospace Council, Inc.

John B. (Brad) Mooney Jr. (NAE) retired from the U.S. Navy with the rank of rear admiral after 34 years of service and is currently an independent consultant in areas relating to ocean engineering and research. His background is in military operations, management, research, education, training, and manpower planning. He has served as president of the Harbor Branch Oceanographic Institution, chief of naval research, oceanographer for the Navy, and Navy deputy of the National Oceanic and Atmospheric Administration. He also served as chair of the NRC Marine Board and as member of the NRC Committee on Autonomous Vehicles in Support of Naval Operations. Admiral Mooney is a member of the National Academy of Engineering.

Judith H. Mopsik is vice president for business development at Abt Associates, Inc. Her areas of expertise include health policy and management; survey instruments and simulations; the veterans system evaluation; and research portfolios at the Department of Defense, Department of Veterans Affairs, and agencies within the Department of Health and Human Services. She has also served as vice president for software applications and biomedical computing at the Computer Sciences Corporation, where her work included systems design and implementation of surveillance and analytic systems for the National Cancer Institute, development of the computation systems of the consumer expenditure survey for the 1987 consumer price index revision, and project management for the large-scale integrated postsecondary education data systems. Previously, Ms. Mopsik served as executive director for strategic market planning at Bell Atlantic.

John E. Rhodes retired from the U.S. Marine Corps with the rank of lieutenant general after 34 years of service and is currently an independent consultant. His background is in development of warfighting concepts and in the integration of all aspects of doctrine, organization, training and education, equipment, and support and facilities to enable the Marine Corps to field combat-ready forces. In his last position he served as commanding general of the U.S. Marine Corps Combat Development Command, where his responsibilities included assessments of current and future operating environments and adaptation of the Corps' training infrastructure and resources in order to ensure that integrated capabilities were delivered to the combatant commanders. General Rhodes has served on numerous scientific advisory committees, including as a member of the NRC Committee on the Role of Experimentation in Building Future Naval Forces.

Nancy T. Tippins is managing principal and senior vice president of Valtera Corporation (formerly Personnel Research Associates). Her areas of expertise include personnel research; strategies associated with employee selection, assessment, and development; equal employment opportunity and affirmative action; and industrial and organizational psychology. She has served as president of the Society for Industrial and Organizational Psychology and is a fellow of the Society for Industrial and Organizational Psychology and the American Psychological Association. Dr. Tippins served as a member of the NRC Committee on Attitudes, Aptitudes, and Aspirations of American Youth: Implications for Military Recruitment.

James L. Wolbarsht, president and chief executive officer of DEFCON, Incorporated, has more than 30 years of experience in business strategy and operations in both the public and private sectors. The firm integrates organizational, operational, and technology tools for commercial and federal clients, including the Department of Defense. Mr. Wolbarsht focuses on corporate strategies that create measurable results and on optimized business processes that lead to effective and lasting change in large, complex organizations. He is frequently cited in the media as a thought leader in the field of business transformation. As the chief financial officer and chief information officer of the Pension Benefit Guaranty Corporation, he was credited with "turning the once-troubled agency around." He has served on multiple scientific boards and committees, including the Pentagon's Ballistic Missile Defense Advisory Committee and the Director's External Review Team at the U.S. National Security Agency. He is a current member of the Naval Research Advisory Committee, the U.S. Naval Services' senior scientific advisory board.

Staff

Charles F. Draper is director of the NRC's Naval Studies Board. Before joining the NRC in 1997, Dr. Draper was the lead mechanical engineer at S.T. Research Corporation, where he provided technical and program management support for satellite Earth station and small satellite design. He received his Ph.D. in mechanical engineering from Vanderbilt University in 1995; his doctoral research was conducted at the Naval Research Laboratory (NRL), where he used an atomic-force microscope to measure the nanomechanical properties of thin-film materials. In parallel with his graduate student duties, Dr. Draper was a mechanical engineer with Geo-Centers, Inc., working on-site at NRL on the development of an underwater X-ray backscattering tomography system used for the nondestructive evaluation of U.S. Navy sonar domes on surface ships.

Arul Mozhi is senior program officer at the NRC's Naval Studies Board. He also served as senior program officer at the NRC Board on Manufacturing and Engineering Design and the National Materials Advisory Board. Prior to joining the NRC in 1999, Dr. Mozhi was senior scientist and program manager at UTRON, Inc., a high-tech company in the Washington, D.C., area, working on pulsed electrical and chemical energy technologies applied to materials processing. From 1989 to 1996 Dr. Mozhi was a senior engineer and task leader at Roy F. Weston, Inc., a leading environmental consulting company working on long-term nuclear materials behavior and systems engineering related to nuclear waste transport, storage, and disposal in support of the U.S. Department of Energy. Before 1989 he was a materials scientist at Marko Materials, Inc., a high-tech firm in the Boston area, working on rapidly solidified materials. He received his M.S. and Ph.D. degrees (the latter in 1986) in materials engineering from Ohio State University and then served as a postdoctoral research associate there. He received his B.S. in metallurgical engineering from the Indian Institute of Technology in 1982.

B

Details on Demographics

DEMOGRAPHIC CHANGES

BOX B.1
Complete Definitions of Population Groups

The following definitions are reproduced from the U.S. Census website:

- White refers to people having origins in any of the original peoples of Europe, the Middle East, or North Africa. It includes people who indicated their race or races as "White" or nationalities such as Irish, German, Italian, Lebanese, Near Easterner, Arab, or Polish.
- Black or African American refers to people having origins in any of the Black racial groups of Africa. It includes people who indicated their race or races as "Black, African-American, or Negro," or nationalities such as Nigerian, or Haitian.
- American Indian and Alaska Native refers to people having origins in any of the original peoples of North and South America (including Central America), and who maintain tribal affiliation or community attachment. It includes people who indicated their race or races by marking this category or writing in their principal or enrolled tribe, such as Rosebud Sioux, Chippewa, or Navajo.
- Asian refers to people having origins in any of the original peoples of the Far East, Southeast Asia, or the Indian subcontinent. It includes people who indicated their race or races as "Asian Indian, Chinese, Filipino, Korean, Japanese, Vietnamese, or other Asian," or wrote in nationalities such as Burmese, Hmong, Pakistani, or Thai.
- Native Hawaiian and Other Pacific Islander refers to people having origins in any of the original peoples of Hawaii, Guam, Samoa, or other Pacific Islands. It includes people who indicated their race or races as "Native Hawaiian, Guamanian or Chamorro, Samoan, or other Pacific Islander," or nationalities such as Tahitian, Mariana Islander, or Chuukese.
- Some other race was included in Census 2000 for respondents who were unable to identify with the five Office of Management and Budget race categories. Respondents who provided write-in entries such as Moroccan, South African, Belizean, or a Hispanic origin (for example, Mexican, Puerto Rican, or Cuban) are included in the "Some other race" category.
- The term Hispanic is an ethnic category **not a racial category** for persons who identify themselves as being of Spanish origin. Unlike other Census Bureau designations, Hispanic denotes neither race nor color, and a Hispanic may be White, Black, or American Indian, as well as (1) Mexican Americans/Chicanos, (2) Puerto Ricans/Boricuas, (3) Hispanos (U.S. Hispanics who identify themselves as Spanish), (4) Cuban Americans, and (5) Latinos (Hispanics from countries other than those already mentioned). Terms other than Hispanic may be preferred. For example, many Mexican Americans prefer Chicano, Puerto Ricans may prefer Boricuas, while others may prefer the more general term, Latino.

NEW WORKFORCE ENTRANTS

TABLE B.1 Breakdown of New Workforce Entrants

	Percentage of Workforce, 1985	Percentage Increase, 1985 to 2000
White male	47	15
White female	36	42
Nonwhite male	5	7
Nonwhite female	5	13
Immigrant males	4	13
Immigrant females	3	9

SOURCE: 2000 U.S. Census (www.census.gov).

TABLE B.2 High School Graduation Rates by Group

	All Races	White	Black	Asian, Pacific Islander	Hispanic
	Male/Female	Male/Female	Male/Female	Male/Female	Male/Female
2000	84.2/84.0	84.8/85.0	78.7/78.3	88.2/83.4	56.6/57.5
2001	84.1/84.2	84.4/85.1	79.2/78.5	90.3/85.1	55.5/58.0
2002	83.8/84.4	84.3/85.2	78.5/78.9	89.5/85.5	56.1/57.9
2003	84.1/85.0	84.5/85.7	79.6/80.3	89.5/86.0	56.3/57.8
2004	84.8/85.4	85.3/86.3	80.4/80.8	88.7/85.0	57.3/59.5
2005	84.9/85.4	85.2/86.2	81.1/81.2	90.4/85.1	58.0/58.9

SOURCE: U.S. Census website (www.census.gov).

ROLE OF THE RESERVE COMPONENT

TABLE B.3 Change in Reserve Component Manning Between 1990 and 2005

Reserve Status and Branch of Service	1990	1995	2000	2005
Total Reserves[a]	1,688,674	1,674,164	1,276,843	1,136,200
Ready Reserve	1,658,707	1,648,388	1,251,452	1,113,427
Army[b]	1,049,579	999,462	725,771	636,355
Navy	240,228	267,356	184,080	140,821
Marine Corps	81,355	103,668	99,855	99,820
Air Force[c]	270,313	263,011	229,009	223,551
Coast Guard	17,232	14,891	12,737	12,880
Standby Reserve	29,967	25,776	25,391	22,773
Army	788	1,128	701	1,668
Navy	11,791	12,707	7,213	4,038
Marine Corps	1,424	216	895	1,129
Air Force	15,369	11,453	16,429	15,897
Coast Guard	595	272	153	41
Retired Reserve	462,371	505,725	573,305	627,424
Army	223,919	259,553	296,004	321,312
Navy	111,961	97,352	109,531	117,093
Marine Corps	9,101	11,319	12,937	14,693
Air Force	117,390	137,501	154,833	174,326

[a]Less Retired Reserve.
[b]Includes Army National Guard.
[c]Includes Air National Guard.

SOURCE: U.S. Department of Defense, DOD Personnel and Procurement Statistics, Personnel, Publications, Atlas/Data Abstract for the United States and Selected Areas, annual; available at <http://siadapp.dmdc.osd.mil/personnel/MMIDHOME.htm>. Last accessed January 7, 2008.

BOX B.2
Breakdown of the Three Reserve Components

All members of a reserve component are assigned to one of three reserve component categories:

1. The **Ready Reserve** comprises military members of the Reserve and National Guard, organized in units or as individuals, liable for recall to active duty to augment the active components in time of war or national emergency. The Ready Reserve consists of the following reserve component subcategories:

— The **Selected Reserve** consists of those units and individuals within the Ready Reserve designated by their respective Services and approved by the Chairman of the Joint Chiefs of Staff as so essential to initial wartime missions that they have priority over all other Reserves. The Selected Reserve consists of additional sub-subcategories:

- **Drilling Reservists in Units** are trained unit members who participate in unit training activities on a part-time basis.
- **Training Pipeline** (nondeployable account) personnel are enlisted members of the Selected Reserve who have not yet completed initial active duty for training (IADT) and officers who are in training for professional categories or in undergraduate flying training.
- **Individual Mobilization Augmentees** (IMAs) are trained individuals assigned to an active component, Selective Service System, or Federal Emergency Management Agency (FEMA) organization's billet that must be filled on or shortly after mobilization. IMAs participate in training activities on a part-time basis with an active component unit in preparation for recall in a mobilization.
- **Active Guard/Reserve** (AGR) are National Guard or Reserve members of the Selected Reserve who are ordered to active duty or full-time National Guard duty for the purpose of organizing, administering, recruiting, instructing, or training the reserve component units.

— **Individual Ready Reserve** (IRR) personnel provide a manpower pool composed principally of individuals having had training, having previously served in an active duty component or in the Selected Reserve, and having some period of their military service obligation (MSO) remaining.

continued

BOX B.2 Continued

— **Inactive National Guard** (ING) are National Guard personnel in an inactive status in the Ready Reserve, not in the Selected Reserve, attached to a specific National Guard unit, who are required to muster once a year with their assigned unit but do not participate in training activities. On mobilization, ING members mobilize with their units.

2. The **Standby Reserve** consists of personnel who maintain their affiliation without being in the Ready Reserve, who have been designated key civilian employees, or who have a temporary hardship or disability. They are not required to perform training and are not part of units but create a pool of trained individuals who could be mobilized if necessary to fill manpower needs in specific skills.

— **Active Status List** are those Standby Reservists temporarily assigned for hardship or other cogent reason; those not having fulfilled their military service obligation or those retained in active status when provided for by law; or those members of Congress and others identified by their employers as "key personnel" and who have been removed from the Ready Reserve because they are critical to the national security in their civilian employment.

— **Inactive Status List** are those Standby Reservists who are not required by law or regulation to remain in an active program and who retain their Reserve affiliation in a nonparticipating status, and those who have skills that may be of possible future use to the Armed Force concerned.

3. The **Retired Reserve** consists of all Reserve officers and enlisted personnel who receive retired pay on the basis of active duty and/or reserve service; all Reserve officers and enlisted personnel who are otherwise eligible for retired pay but have not reached age 60, who have not elected discharge, and are not voluntary members of the Ready or Standby Reserve; and other retired reservists under certain conditions.

SOURCE: Available at <http://en.wikipedia.org/wiki/Reserve_Component_of_the_Armed_Forces_of_the_United_States>. Accessed on October 5, 2007.

BOX B.3
Mobilization Categories

- **Full Mobilization** requires a declaration of war or national emergency by the Congress, affects all reservists (including those on inactive status and retired members), and may last until six months after the war or emergency for which it was declared.
- **Partial Mobilization** requires a declaration of national emergency, affects only the Ready Reserve, and is limited to a maximum of one million personnel activated for no more than two years.
- **Presidential Reserve Call-Ups** do not require a declaration of national emergency but require the President to notify Congress, and each is limited to 200,000 Selected Reservists and 30,000 Individual Ready Reservists for up to 270 days.
- The **15-Day Statute** allows individual service secretaries to call up the Ready Reserves for up to 15 days per year for annual training or operational missions.
- **RC Volunteers** may request to go on active duty regardless of their reserve component category, but the state governors must approve activating National Guard personnel.

SOURCE: Available at <http://en.wikipedia.org/wiki/Reserve_Component_of_the_Armed_Forces_of_the_United_States>. Accessed on October 5, 2007.

C

Acronyms and Abbreviations

AFQT	Armed Forces Qualification Test
AIP	assignment incentive pay
ARCI	acoustic rapid COTS insertion
ASN(M&RA)	Assistant Secretary of the Navy for Manpower and Reserve Affairs
ASVAB	Armed Services Vocational Aptitude Battery
ASW	antisubmarine warfare
AVF	all-volunteer force
BAH	basic housing allowance
BAS	basic allowance for subsistence
BRAC	Base Realignment and Closure
CAOCL	Center for Advanced Operational Cultural Learning
CARS	Country, Area, Regional Specialist
CARSO	Country, Area, Regional Specialty Officer
CBO	Congressional Budget Office
CMC	Commandant of the Marine Corps
CNA	Center for Naval Analyses
CNO	Chief of Naval Operations
COTS	commercial off-the-shelf
DACMC	Defense Advisory Committee on Military Compensation
DACOWITS	Defense Advisory Committee on Women in the Service

DASN(TFT)	Deputy Assistant Secretary of the Navy, Total Force Transformation
DCMC(M&RA)	Deputy Commandant of the Marine Corps for Manpower and Reserve Affairs
DDG	guided missile destroyer
DDX	next-generation destroyer
DOD	Department of Defense
DOPMA	Defense Officer Personnel Management Act
EOD	explosive ordnance disposal
EW	electronic warfare
FAO	foreign area officer
FMTU	Foreign Military Training Unit
FSA	family separation allowance
GAO	Government Accountability Office
GWOT	global war on terror
HCS	human capital strategy
HSV	high-speed vessel
IA	individual augmentee
IED	improvised explosive device
IRR	Individual Ready Reserve
IT	information technology
JAMRS	Joint Advertising, Market Research and Studies
JASS	Job Assignment Selection System
JROTC	Junior ROTC
LCS	littoral combat ship
MPT&E	Manpower, Personnel, Training, and Education
NAVAIR	Naval Air Systems Command
NCO	noncommissioned officer
NRC	National Research Council
NJROTC	Naval Junior Reserve Officer Training Corps
OJT	on-the-job training

PC	patrol coastal (ship)
PPO	preferred provider organization
PSYOP	psychological operation
QoL	quality of life
RAO	Regional Area Officer
RC	Reserve Component
RECON	reconnaissance
Redux	Military Retirement Reform Act
RF	radio frequency
RMC	regular military compensation
ROPMA	Reserve Officer Personnel Management Act
ROTC	Reserve Officer Training Corps
SDIC-C	Sea Duty Incentive Compensation, Curtail
SDIC-E	Sea Duty Incentive Compensation, Extend
SEAL	sea, air, land (team)
SECNAV	Secretary of the Navy
SRB	selective reenlistment bonus
SSB	selective separation benefit
TSP	Thrift Savings Plan
USAID	U.S. Agency for International Development
USERRA	Uniformed Services Employment and Reemployment Rights Act
VBSS	visit, boarding, search, and seizure
VLS	vertical launch system
VSI	voluntary separation incentive
VSP	voluntary separation pay
WOMA	Warrant Officer Management Act
WWII	World War II
YATS	Youth Attitude Tracking Study
YOS	years of service